HOMEMADE GUNS
AND
HOMEMADE AMMO

by Ronald B. Brown

This book is sold for informational purposes only. Neither the author nor the publisher will be held accountable for the use or misuse of the information contained in this book.

Homemade Guns and Homemade Ammo
© 1986 by Ronald B. Brown
© 1999 by Breakout Productions, Inc.

Published by:
Breakout Productions, Inc.
PO Box 1643
Port Townsend, WA 98368

ISBN 1-893626-11-3
Library of Congress Card Catalog 99-60234

WARNING!

The information contained in this book is potentially harmful. In addition, home refined chemicals are not in their purest form and this could cause instability and erratic performance. Also, it may or may not be legal to manufacture the articles described.

DISCLAIMER

This book in no way condones illegal activity! It is *your* responsibility to determine the legality of your actions. Further, because I have no control over the workmanship, materials, tools, methods, or testing procedures employed, I hereby disclaim any responsibility for consequences resulting from the fabrication or compounding of any item described in this book. I cannot and will not accept any responsibility for this information and its subsequent use. This book is sold for informational purposes only!

ACKNOWLEDGEMENTS

Special thanks are in order to *SURVIVE* magazine (now *GUNS & ACTION)* for permission to reprint in Chapter 5 information previously published by me (January-February 1983) in an article entitled *"Strike Anywhere Ammo."*

Thanks also to Mr. Turner Kirkland of Dixie Gunworks for the many muzzle loading tips appearing in his 500 page Dixie Gunworks' catalog.

The information on caliber capabilities in Chapter 2 was drawn from **Speer's Reloading Manual Number Ten;** from *Outdoor Life's* **Gun Data Book** by F. Philip Rice; and from "American Bulleted Cartridges" by Ken Waters, a regular feature of *Gun Digest* for many years.

I discovered the formula for recoil, given in Chapter 2, in the August 1980 issue of *Ammo.* Its appeared in an article entitled *"Recoil: Its Causes, Effects, and Remedies"* by Bob Zwirz.

Thanks also to the Information Publishing Co. (PO Box 10042, Odessa, TX 79767-0042) and Mr. Tim Lewis who wrote **Kitchen Improvised Plastic Explosives.** It contained the idea of obtaining potassium chlorate from bleach and sodium chlorate from table salt via electrolysis, elaborated upon in Chapter 6 of this work.

A similar thanks goes to Atlan Formularies and Mr. Kurt Saxon who wrote **The Poor Man's James Bond.** In it, he suggests reclaiming potassium chlorate from matches.

TABLE OF CONTENTS

CHAPTER 1 The Need For This Book 1

Rationale — Legal Considerations

CHAPTER 2 A Mini-course in Ballistics 8

Rifles — Shotguns — Flintlocks — Percussions — Center Fire — Rimfire — Break Action — Rolling Blocks — Bolt Actions — Lever Actions — Pumps — Automatics — Head Space — Trajectory — Caliber Explanation — Caliber Capabilities — Gauge Explanation — Choke — Recoil — Stopping Power

CHAPTER 3 The Basic Single Shot From Pipe 34

Making the Individual Parts — Assembly — Remote Test Firing

CHAPTER 4 Variations 84

Pipe Dimensions — Straight-sided Shells with Rims — Straight-sided, Rimless Shells — Retaining Rings — Necked Cartridges — Reducers — A Double Barrel — A Muzzle Loader — Fuses — A Wooden Gun

CHAPTER 5 Gunpowder 111

How It Works — Black Powder —
Crushed Match Heads — Performance
Rating — Potassium Chlorate & Sugar —
Sodium Chlorate & Sugar — Potassium
Perchlorate & Sugar — Saltpeter & Sugar
— Other Oxidizers — Making & Buying
Saltpeter — Electrolysis

CHAPTER 6 Primers 131

How They Work — Strike Anywhere
Match Tips — Sulfur & Potassium Chlor-
ate —Mercury Fulminate — Sources of
Materials — Potassium Chlorate From
Match Heads — Potassium Chlorate
From Bleach — Softening Work Harden-
ed Primers — Misfires — SAFETY!

CHAPTER 7 Kitchen Sink Reloading 161

.22 Rimfire — Centerfire Cartridges —
Loads — Bullets — Steps in Reloading —
Shotgun Shells — Patterning

BIBLIOGRAPHY 179

1

THE NEED FOR THIS BOOK

This book will tell you how to make guns, gunpowder, and primers from common materials. No knowledge of chemistry is necessary. Nor is a fancy machine shop. Ordinary hand tools will do.

With this book in hand, anyone who wants firearms can have them — with no litmus test of political, religious, or ethnic worthiness required. There will be people, I am sure, who will question the wisdom of making this information so readily available. The next few paragraphs address these concerns.

My justification has three main propositions. First, firearms are a kind of technology. Second, technology is the great equalizer, the great liberator. Third, technological development goes through a cycle in which it is first an enslaver. Only later does it become a liberator.

Let me illustrate. Today, a woman argues with her husband and kicks him out. It's cold, so she turns up the thermostat. In contrast, the women of a hundred years ago would have faced not a thermostat, but a stack of firewood and a splitting maul.

The husband today is physically stronger than his wife, but so what? A hundred years ago the woman would have needed the man to split the firewood. She is no longer dependent. What has equalized them? The thermostat. The furnace. Technology.

Today's woman may deceive herself into thinking that she is wiser or more enlightened than her grandmother. I submit that *technology* makes the modern woman independent, not enlightenment. And not law. There must be a technological base supporting the law. The Equal Rights Amendment would not be enforceable in a stone age culture, but would be in a hi-tech society.

When a new technology is first discovered, it is known to only a few people and thus puts power in the hands of only a few. In the short run, technology enslaves. Later on, because you can't effectively legislate against discovery, or prevent forever the spread of ideas, the now-old technology is in the hands of many. Power is in the hands of many. And in the long run, all men (and women) are more equal, more free, because of it.

Firearm technology brought an ultimate end to the class system in Europe. But it first went through an enslaver phase where the power represented by the technology was controlled by only a few people.

For example, the Peasants' Rebellion, associated with Martin Luther, occurred nearly 500 years ago. (Luther, you may recall, started the Protestant Reformation.) Gunpowder and firearms came to the field of battle during his lifetime. During the Peasants' Rebellion, the German princes turned their armies, with firearms, upon the unarmed peasants.

The rebellion was put down (in the short run) and serfdom continued. Peasants were slaughtered by the thousands. In Calmar, near Alsace-Lorraine, one can still see today, 500 years later, a huge burial mound where peasant bodies were heaped up to rot.

Over the following three centuries, firearm technology spread — as did technology of all kinds — and men slowly became free. Slavery and serfdom were out. Majority rule was in. The basis of the change was technology.

Another example of weapons technology enslaving people is the Spanish conquest of Peru. The conqueror Pizarro had 168 soldiers (including 67 on horseback), steel blades all round, less than 20 crossbows, and three matchlock firearms. With this rag-tag band of fortune seekers he conquered a nation of millions — an organized nation with a standing army and roads and temples and tax collectors and state-run agriculture. But a nation without steel blades or firearms. This is what happens when technology is controlled by a few.

The other side of the coin was the American Revolution. In that conflict, everyone had guns: farmers and hunters and trappers and explorers and pioneers. Technology was in the hands of many. When the masses have physical power equal to the rulers, autocracy cannot survive.

At some level of consciousness, the framers of the Constitution realized that their new-found independence was rooted in technology, especially firearm technology. They tried to guarantee the continuance of that technology with the Second Amendment.

The right to bear arms has, of course, been infringed in countless ways. (It's amazing what contradictions the legal mind can rationalize.) But the intent was there. The founders of this country knew from whence came their freedom. It was not their ideals or their religion. After all, serfdom had existed for centuries in Christian Europe. It was firearms technology that made them free, not religion.

When technology is in the hands of a few, an army of thousands can control a population of millions. Consider today's situation in Poland, Chile, South Africa, or the Philippines.

What if the information in this book was known by the blacks of South Africa? By the Poles? By the Jews in Nazi Germany? By the citizens of El Salvador? Would the death squads prowl so freely?

I think not. I believe that if every human being on earth understood the contents of this book, the forces of oppression, which now tread so heavily in various parts of the world, would walk with a far lighter footfall.

3

Legal Considerations

The question arises of legality. Is it legal to build your own firearm? To make your own gunpowder? My original intent when compiling this book was to gather up the pertinent laws on the topic and present them for the reader's benefit.

A bit of legal research reveals that it's not that simple. It also reveals that what the average taxpayer thinks about his right to bear arms is not at all what the legal profession thinks. The following is an excerpt from *American Jurisprudence* — an encyclopedia of law written by lawyers for lawyers. It is a standard reference book.

Volume 79, American Jurisprudence 2nd Edition, Weapons and Firearms, Section 4: "No absolute right to keep and bear arms was recognized by common law. Furthermore, the Second Amendment of the Constitution of the United States in declaring that the right of the people to keep and bear arms shall not be infringed means no more than that this right shall not be infringed by Congress, and the guaranty of this Amendment is not carried over into the Fourteenth Amendment so as to be applicable to the states.* The right to bear arms does not apply to private citizens as an individual right guaranteed by the Constitution of the United States. Accordingly, it is generally recognized that state or municipal regulation of weapons does not per se offend the right to bear arms guaranteed by the United States Constitution. Furthermore, the registration provisions of the National Firearms Act cannot be considered an infringement of the federal constitutional guaranty of the right to keep and bear arms."

The Fourteenth Amendment, referred to in this passage, had to do with giving freed slaves full citizenship status after the Civil War. It says, in part, "No State shall make or enforce any law which shall abridge the privileges or immunities of citizens of the United States..."

American Jurisprudence is not casually written. The flow of ideas from sentence to sentence and from paragraph to paragraph makes sense, but every sentence is footnoted and represents a synopsis of a significant court case. Even phrases within sentences are footnoted. No freestanding sentences are thrown in to make the connection from one idea to the next.

Although every sentence in *American Jurisprudence* is footnoted, I am citing below just one (marked with an asterisk in the above quotation):

Harris v. State, 83 Nev 404, 432 P2d 929, 30 ALR 3d 1412.

The Second Amendment was not adopted with the individual's rights in mind, but as a protection for the states in the maintenance of their militia organization against possible encroachments by the federal power. Burton v. Sills, 53 NJ 86, 248 A2d 521, 28 ALR 3d 829, app dismd 394 US 812, 22L Ed 2d 748, 89S Ct 1486.

I must admit, it jarred me to find the above quotes. I'm probably more provincial than I care to believe. As a boy, I attended a one room schoolhouse and was steeped in rural American values. Later in life, researching the family tree, I found seven ancestors who fought as patriots in the American Revolution. They probably thought they were fighting to guarantee me some basic human rights. They must be spinning in their graves.

There are three problems in my reciting for you the legalities of making your own firearms. First, the same laws are interpreted differently by different people. (Note the above discussion on the Second Amendment.) Second, the law may

5

change by the time you read this. And third, various bureaus can make rulings which have the force of law.

The Code of Federal Regulations (27 CFR 178.11) defines *manufacturer* as "Any person engaged in the manufacture of firearms or ammunition for purposes of sale or distribution." In Section 178.41 it says that a manufacturer must be licensed.

Over the ages, the courts have adopted the doctrine of "strict interpretation" in regard to criminal law. This means that a person should not be held accountable for a criminal act without having notice that the act was, indeed, criminal. The only way to insure adequate notice is to interpret the law quite literally.

If you add all this together, it means you don't need a license to make a firearm if you don't intend to sell or distribute it. I have seen this interpreted as meaning that it's legal to make your own gun.

There are kits on the market with which you can build your own muzzle loader. In the absence of a law making these kits illegal, it appears by default that they *are* legal. You can also buy substitute and replacement parts for these kits. What's the difference between substituting part for part and simply building your own from scratch? What's the difference between building a kit and "gunsmithing" (which requires a license)?

I'm sure the answer to some of these questions is in the law library at your local court house. And I'm equally sure that the answer to some questions is not there — no matter how long you look.

Another complicating factor is the Bureau of Alcohol, Tobacco, and Firearms (BATF) and the role it plays in administering the National Firearms Act (NFA).

The NFA was enacted in 1934 as a tax law. The NFA makes it illegal to possess, without a special license, certain kinds of firearms — machine guns, sawed-off shotguns, etc. These are referred to as "NFA weapons." The National Firearms Act also provided that the BATF would administer these laws. It is up to the director of the BATF to decide what constitutes an NFA weapon. If the interpretation changes, then what is a legal or illegal activity changes.

The same kinds of questions and problems exist in relation to making gunpowder. Gunpowder itself is sold without restriction in sporting goods stores. The ingredients with which to make it are also sold without restriction. All this implies that it would be legal to make your own.

Let's say that you, the reader, want to know if it's legal to make your own gunpowder. Obviously, if you find a law which forbids it, then it's not legal. But if you don't find one and yet there is such a law, you are still bound by it whether you find it or not.

The dilemma is this: If you can't find one, how do you know if there is one? Further, if you do find one, how do you know that it's the most *recent* one? These are very thorny questions.

Unfortunately, my own experience is that the more time I spend digging in the law books, the more questions I come up with instead of answers. To pursue it further (and I heartily recommend that you do), ***Legal Research*** by Stephen Elias (Nolo Press, 1982) is an excellent place to start. May I say that you have both my heartfelt blessings and condolences in advance.

2

A MINI-COURSE IN BALLISTICS

If you already understand firearms, you may wish to skip this chapter. If, however, you are hiding in an attic in Nazi Germany and you know *nothing* of firearms, this chapter might be very useful. When you read or hear "single shot" or "trajectory" or "rifling" you need to know what it means. The purpose of this chapter is to provide a basic grasp of firearm terms.

Rifles and Shotguns

In simplest terms, a firearm consists of a barrel, a projectile, and a charge of gunpowder. The projectile (bullet) flies through the air and hits the target. The barrel aims the projectile — guides it — and traps the explosive force of the gunpowder behind the bullet. Gunpowder provides the driving force for the bullet.

In very early guns, the barrel was simply a metal tube, closed at one end to trap the explosive force of the gunpowder and open at the other to let the bullet escape. The inside surface of the tube was smooth. Such guns had limited accuracy by today's standards.

An understanding of archery helps in understanding firearms. The feathers on the back of an arrow are not attached parallel to the shaft. Rather, they are mounted at an angle. When the arrow is shot, air rushing past the angled feathers forces the arrow to spin while in flight, similar to a football spiraling about its lateral axis. A spiraling football, or arrow, flies truer to the mark than does a non-spiraling projectile.

How to make the bullet spin? The answer is in "rifling" — spiral grooves cut into the inside of the gun barrel. The grooves bite into the soft lead bullet and give it a twist before it exits the muzzle.

Guns with rifling cut into the barrels are called "rifles." Guns with barrels which are smooth on the inside, where no rifling has been cut, were once called "smoothbores." The only smoothbores made today are "shotguns." This name implies the use of "shot" — small pellets instead of a single projectile.

Factory-made shotguns, loaded with shot which rapidly disperses into a large diameter circle, will carry with killing power only to 50 yards. Shotguns loaded with slugs — a single lump of lead instead of pellets — are accurate (at best) to only a hundred yards. For accuracy, rifles begin where shotguns leave off. Ordinary hunting rifles are accurate to over a hundred yards, some calibers out to 300 yards, and specialized rifles and sights out to a thousand yards.

Muzzle Loaders

Firearms are also classified by the load-and-feed mechanism used for repeat shots.

A muzzle loader is a gun which is loaded from the muzzle. The gunpowder is carried loose in a separate container and poured down the muzzle. The bullet is then pushed down the muzzle also, and seated on the powder charge. Loading a muzzle loader is a fairly time consuming affair. It takes a skilled shooter some 15 seconds to get off a second shot.

In a muzzle loader, one end of the barrel is sealed off with a "breech plug." Gunpowder is trapped between the breech plug and the bullet. Two different systems are used to introduce fire to the gunpowder — the flintlock system and the percussion cap system.

With a "flintlock," a piece of flint stone is clamped in a spring-loaded holder. When the trigger is pulled, the flint strikes against a piece of steel and a shower of sparks falls onto a small pan which holds a pinch of gunpowder.

The powder ignites and the flame follows a gunpowder path through a tiny hole drilled in the wall of the barrel. This is the way in which the main powder charge inside the barrel is ignited. The firing sequence itself takes a noticeable amount of time. The trigger is pulled, some fizzing and sputtering occurs, and then the gun goes bang.

An improvement on the flintlock was the "percussion" ignition system. With a percussion gun, fire is more reliably introduced to the main charge. The percussion "cap" is a small metal foil cup which contains a primary explosive (that is, an explosive which is sensitive to friction and shock as well as spark).

A percussion cap is mounted on a "nipple" and struck with a spring-loaded hammer. The hammer is released by pulling the trigger. The percussion cap "explodes" and produces a hot flame. The nipple resembles a grease fitting. It is threaded on the outside and screws into the barrel. It is hollow on the inside and carries the flame from the percussion cap to the powder charge.

Percussion guns have more reliable ignition than flintlocks but, because they both load through the muzzle, they are equally slow in reloading.

Cartridges

With a muzzle loader, the component parts of a charge were carried separately and loaded separately into the gun. When cartridges evolved, they represented a unitized charge where all the components (bullet, gunpowder, and primer) were preloaded into a brass shell.

Using a preloaded shell, repeat shots could be accomplished much faster than with muzzle loaders. Mechanisms were invented which could even introduce the next shell into the gun's firing chamber with no manipulation at all being required on the part of the operator.

"Center fire" cartridges use a primer which is very similar to the old-time percussion cap. The primer looks like a dot or button on the rear of the cartridge. When the trigger is pulled, the gun's firing pin strikes the primer which contains a small amount of primary explosive and ignites the main powder charge within the shell.

A "rim fire" cartridge contains no primer cap. Instead, the primary explosive is contained in the hollow rim of the brass shell. The gun's firing pin crushes the rim, igniting the primary explosive, and detonating the main powder charge.

The only rimfire guns made today are .22 caliber rifles, suitable for small game. For all practical purposes, rimfire shells cannot be reloaded. The shell itself is damaged in firing.

All other rifle calibers and all shotgun shells are centerfire. Centerfire shells can be reloaded because only the primer is dented in firing, not the shell casing. The spent primer can be pressed out and replaced with a fresh one.

Some shells have "rims." Others do not. See Figure 2-1. Rims and lack of rims and grooves in the shell casing are incorporated in the cartridge design in order to accommodate the gun's extraction and repeat shot mechanism.

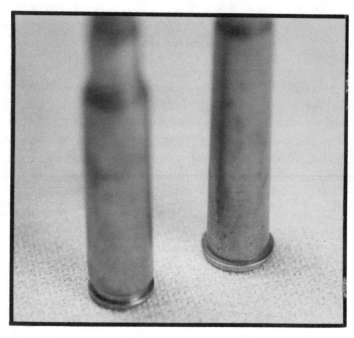

Figure 2-1

The cartridge on the right has a rim. The one on the left does not and is called "rimless." Both are .30 caliber shells.

Single Shot Firearms

In the early days, single shot guns were called "breechloaders" — because shells were inserted in the rear of the gun — to distinguish them from muzzle loaders. The break action, the rolling block, and the bolt action without repeat shot capability are the basic design types of single shots.

Break actions are hinged in the middle. See Figure 2-2. They "break" in half, allowing the cartridge to be loaded into the breech end of the barrel. When closed, the cartridge is trapped inside the gun. This is a very strong action. High powered African big game rifles are very often break action guns.

Figure 2-2

Two guns, both of them break actions. On the left is a single shot; on the right, a side-by-side double. Over-and-under double barrels are also made. They, too, are break action.

The rolling block is a breech loading single shot. See Figure 2-3. With a rolling block, the steel lug which blocks the breech end of the barrel retracts into the stock when a lever is activated. The lever also serves as a trigger guard. This is a strong action.

The same Strong, Rugged Action as the Ruger No. 1 Rifle...

...with an American style lever.

Figure 2-3
The rolling block action.

Bolt Actions

The bolt action (see Figure 2-4) can offer repeat shot capability as well as single shot capability. If a spring-loaded clip is used to hold the follow-up shells, not only is the old, fired shell extracted from the gun when the bolt is opened, a new round is automatically inserted in the chamber when the bolt is closed. The bolt action is the slowest of the repeating action designs but is also the strongest, most jam-free, and most accurate. Sniper rifles are bolt action.

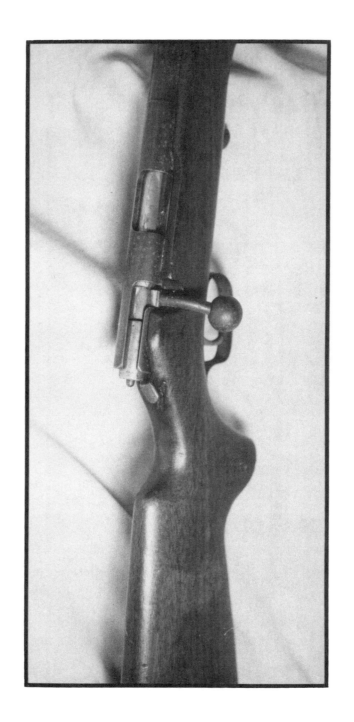

Figure 2-4
The bolt action.

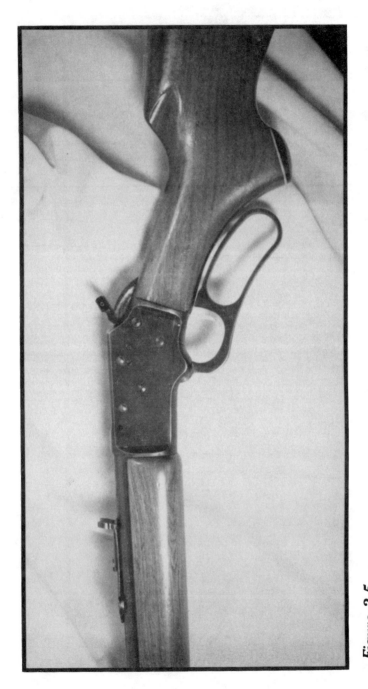

Figure 2-5
The lever action.

Lever Actions

The lever action is the saddle gun used by cowboys in Western movies. It is flat, with no protrusions, and slides easily in and out of a saddle scabbard. The lever action is not a strong action and it is chambered in only the milder calibers.

If the action is operated calmly and smoothly, repeat shots are very fast. In haste or panic, however, jams are common, regardless of brand name or caliber. See Figure 2-5.

Pumps

With a pump action, the foregrip portion of the stock slides back and forth like a slide trombone. After firing, pulling back on the sliding foregrip ejects the spent shell. Pushing forward loads the next shell into the chamber.

Pump actions have faster repeat shot capability than any other design described so far, are cheaper than semi-automatics, and are more jam-free. They have the disadvantages of poor accuracy (the foregrip tends to wobble as you aim) and of being noisy — the foregrip rattles, and this is not a great blessing to the still hunter or stalker. See Figure 2-6.

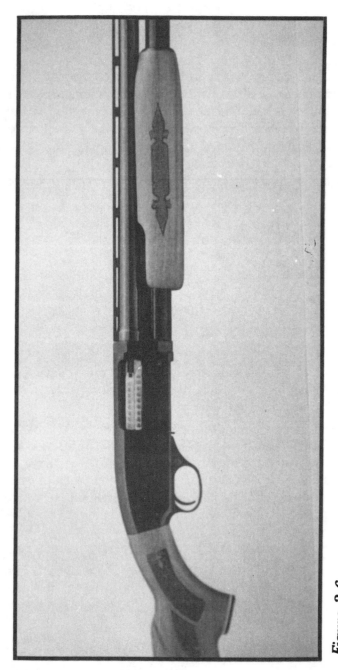

Figure 2-6
The pump action.

Automatics

An automatic rifle is one which ejects the old, spent cartridge and then inserts a new, fresh cartridge in the firing chamber with no manipulation required on the part of the shooter. We think of automatic weapons as modern, but the first automatic weapon used by a military organization was employed by the Danish navy in the 1880's — a hundred years ago!

With a "semi-automatic" rifle, the trigger must be pulled, released, and pulled again for each repeat shot. With a "full automatic" rifle, the gun continues to fire, round after round, as long as you hold back the trigger. In everyday language, both types are called "automatics."

With other action types, the power required to eject the old shell and insert the new shell is provided by the arm of the shooter. With an automatic, the power required to work the action is derived by stealing some of the force being used to propel the bullet.

A hole is drilled sideways into the rifle barrel. When the gun is fired and the bullet is halfway to the muzzle it passes this hole and some gas escapes. The pressure of the gas escaping through the hole is channeled to throw the bolt — otherwise done by the arm of the shooter.

The advantage of an automatic weapon is fast repeat shots with no distraction to the shooter. He can hold his aim on the target and continue to shoot. The disadvantages are cost (automatics are expensive), reliability (automatics are complicated mechanisms and sometimes jam), and strength (the most powerful calibers are not chambered in automatics). See Figure 2-7.

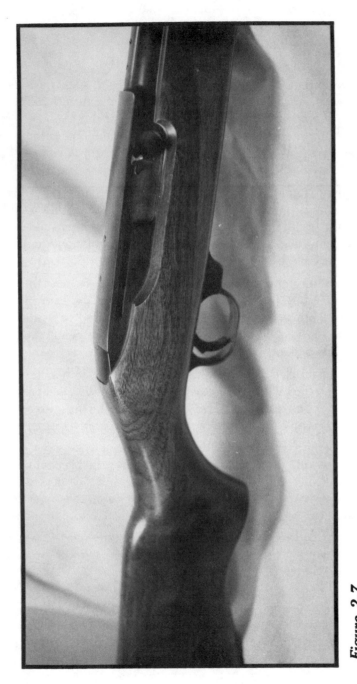

Figure 2-7
The semi-automatic.

Head Space

In a factory-made gun, the space which exists between the rear of the cartridge and the front of the breech plug is called "head space." If the head space is excessive, the expanding gunpowder inside the cartridge stretches the cartridge case until it breaks. If this happens, damage to the gun usually results — and sometimes to the shooter, as well.

Head space increases as a gun wears. Gunsmiths have dummy cartridges they can use to check the head space in old guns. Past a certain point, too much headspace means that a gun is not safe to shoot.

To understand why, think of a prizefighter punching his opponent in the jaw. Imagine his glove being glued to the other man's jaw. He couldn't get off much of a punch if no slack existed. But if his fist had two inches of "running start," he could do some real damage.

The same principle holds true in firearms. If the cartridge fits loosely in the gun's chamber, the surge of pressure can rupture the shell casing and give it a running start before it slams into the rear wall of the chamber. In the homemade gun described in Chapter 3, the breech plug screws into the rear of the barrel and eliminates head space. It automatically compensates for differences in rim thickness, wear in the gun, and so forth.

A different kind of head space can exist in homemade firearms and be a cause for concern, however. The mismatch of diameters between the outside of the cartridge and the inside of the barrel sometimes results in a sloppy fit. The best policy is to test fire the gun by remote firing (see Chapter 3, Figure 3-50) prior to hand-held firing. It should be test fired with a more powerful load than will be employed in everyday use.

Trajectory

If you roll a marble across a table top, as soon as it rolls off the edge it starts falling toward the floor. The faster you roll it, the further away from the table it lands. Nevertheless, as soon as it is airborne, gravity starts acting upon it.

The same thing is true of a bullet leaving a gun barrel. As soon as the bullet is airborne, as soon as nothing is under it holding it up, it begins to fall. How far will it travel before it hits the ground? That is the gun's range. The range can be extended by aiming the gun slightly upwards.

If you aim the gun slightly upwards, the bullet will travel in a long arc. At point blank range the bullet will hit the center of a bull's-eye. At a short distance, the bullet will hit above the bull's-eye. This is because you have aimed the gun upward slightly to begin with.

The bullet will finally reach the top of its arc and begin to fall. At some distance from the target, the bullet will again strike the center of the bull's-eye. It is now falling to earth. As the distance from the gun's muzzle to the target increases past this point, each successive bullet will strike lower and lower on the target. This phenomenon is called "trajectory."

Caliber Explanation

The topic of calibers is confusing. You might think that the bigger the caliber, the more powerful the gun. 'Taint necessarily so. Although the caliber does denote the bullet diameter, you can have a small bullet backed up by a large powder charge or a large bullet backed up by a relatively small charge. The following is a basic list of rifle calibers and a description of their capabilities.

Rifles

.22 Rimfire .22 shorts are good to 50 yards on small pests such as rats, squirrels, crows, and snakes. .22 long rifle cartridges are adequate to 75 yards. .22 shorts can be stabilized with a 1-24 rifling twist. .22 long rifles require a 1-16 twist. The longer the bullet, the sharper must be the twist.

.22 Hornet Introduced by Winchester in 1930. Good to 175 yards on varmint-sized game (foxes, turkeys). It is too light a caliber for deer. The metric designation is 5.6 x 35R. The rifling twist is 1-16 (that is, one full twist for each sixteen inches of barrel length).

.222 Remington Introduced in 1950. Good to 225 yards on varmints. Outstanding accuracy to 200 yards. Rifling twist is 1-14.

.223 Remington Adopted by the U.S. military in 1964 as the 5.56 mm. An excellent varmint cartridge to 250 yards. Metric designation is 5.56 x 45mm. Rifling twist is 1-14.

.22-250 Began as a wildcat cartridge in the 1930's. Wildcat means home-loaded; not available in factory loads. It was a .250 Savage case necked down to accept a .22 caliber bullet. Now available commercially, it's a varmint cartridge with an effective range to 300 yards. Rifling twist is 1-14. This is a high velocity, flat shooting load with a loud muzzle blast. The supercharged load results in fast barrel wear.

.220 Swift Introduced in 1935 by Winchester. Effective range to 350 yards. It was the first factory cartridge with a muzzle velocity of over 4000 feet per second (fps). Rifling twist is 1-14. No rifle is made for the .220 Swift today.

.243 Winchester This is really a .308 Winchester necked down to take a .243" diameter bullet (6mm). In the wind, it is superior to any .22 caliber bullet. Good to 300 yards; big enough for deer. Light recoil. Rifling twist 1-9. Metric designation is 6 x 51mm.

.250 Savage Introduced by Savage Arms in 1915. Delivered 3000 fps muzzle velocity. Good to 250 yards. Popularity declined with the introduction of the .243 Winchester.

.257 Roberts Designed by Ned Roberts and introduced by Remington in 1934. Good to 250 yards and big enough for deer. Rifling twist of 1-10. Popularity waned with the introduction of the .243 Winchester.

.270 Winchester The .270 is a necked-down .30-06 case. Good to 325 yards. With hand loads, it ranks near the top for long range accuracy. With heavier bullet weights it will take all North American big game except the larger bears. Rifling twist is 1-10. A contender for best all-round big game rifle.

7mm Remington Magnum Introduced in 1962. Very flat trajectory. Effective range of 350 yards. Rifling twist is 1-9. Fast barrel wear results from "hot" loads.

7mm Mauser The Spanish military cartridge of 1893. Velocity and trajectory characteristics are less than the .270 but killing power is far in excess of the .30-30.

.30 Carbine The .30 M1 Carbine was developed by the Army in 1940 as a substitute for a pistol. As a rifle cartridge it is neither very powerful nor accurate. It should not be used on game larger than coyote. Useful range is 100 yards.

.30-30 Introduced in 1895 for use in the '94 Winchester lever action. It was the first sporting cartridge loaded with smokeless powder. Its popularity is due to the guns which are chambered for it — flat, light lever actions that are well balanced and easy to handle. Adequate on deer to 175 yards. The metric designation is 7.62 x 51R. Rifling twist is 1-12.

.300 Savage Introduced in 1920 for use in the Savage 99 lever action. It has nearly the power of a .30-06 but has a much shorter case, thus lending itself to action types other than bolt action. Declined in popularity with the introduction of the .308. Useful range is something less than 300 yards.

.308 Winchester Civilian version of the 7.62mm NATO cartridge. Better performance than a .300 Savage and nearly as good as a .30-06. It has the advantage to gun designers of having a shorter case than the .30-06. Rifling twist is 1-10.

.30-06 Springfield A .30 caliber cartridge adopted by the U.S. Army in 1906. Generous case capacity makes it a versatile cartridge for reloading. Useful range to 325 yards. Adequate for all North American big game except grizzlies and Alaskan brown bear. Rifling twist is 1-10.

.300 Winchester Magnum A belted magnum cartridge introduced in 1956. Has flat trajectory. When sighted in at 250 yards, is 2.9" high at 150 yards and 3.5" low at 300 yards. The 7mm magnum is slightly flatter but not as powerful. Special target rifles have produced outstanding accuracy to 1000 yards. Useful range for hunting arms is 350 yards. Adequate for all North American big game except the larger bear. Rifling twist is 1-10.

.303 British The British service cartridge adopted in 1888 and used through Korea. The shell is both rimmed and tapered. The bullet is .30 caliber, but the rear end of the case is .455 inches, only .001 less than a .44 Magnum. This is a long cartridge like the .30-06 with ballistics similar to the .300 Savage. Metric designation is 7.7 x 57R.

.35 Remington Introduced in 1906. A competitor of the .30-30. Has better stopping power than the .30-30. The rifling is typically cut shallow, wears comparatively quickly, and accuracy suffers.

.375 Winchester A recent introduction by Winchester. A competitor of the .444 Marlin.

.375 H & H Magnum Introduced in 1912 by Holland & Holland. An African big game rifle, unnecessarily powerful for North American big game. Unique in that it shoots different bullet weights to the same point of impact. Double the recoil of the .30-06.

.44 Magnum Introduced as a pistol cartridge in 1956, now chambered in rifles. In the same class as a .30-30 or .35 Remington, it has better stopping power than either.

.444 Marlin Introduced in 1964 for Marlin's lever action. Useful range is 150 yards. Can be thought of as a much beefed-up .44 Magnum, from a performance point of view.

.45-70 Government U.S. military cartridge from 1873 to 1892. It has a straight case and is rimmed. Originally loaded with black powder, it has a large case and can be hand loaded with 500 grain bullets to a point where it approaches a .458

Winchester. Only new single shot and bolt action rifles are strong enough for this. The old "trap door" military rifles would burst from such pressures. With factory loads, effective range is 150 yards.

.458 Winchester Magnum Introduced in 1956. Replaced the .375 H & H as the most powerful commercial load. Also has the most recoil. Jacketed, 500 grain bullets are available for rhinos and tanks. Rifling twist is 1-14. The cartridges are belted.

Pistols

9mm Parabellum This is the 9mm Luger cartridge adopted by the German military in the early 1900's. It is used in both automatic handguns and submachine guns, not for serious target competition. It is not a powerful load. The shell does not have a rim.

.38 S&W Introduced in 1877 by Smith & Wesson. Slightly less powerful than the .38 Special. A straight, rimmed case.

.38 Special During the 1899 Philippine campaign the then-standard .38 Long Colt was found inadequate. The .38 Special replaced it and is the most widely used pistol cartridge of all time. Adequate for varmint hunting. Straight, rimmed case.

.357 Magnum Dates from 1935. Triple the muzzle energy of the .38 Special. Straight, rimmed case. It is chambered in some rifles.

.41 Magnum Between the .357 Magnum and the .44 Magnum in both performance and recoil.

.44 Magnum See rifle cartridges, above.

.45 Auto Rim Dates from 1922. Reduced popularity since the introduction of the .44 Magnum. Straight, rimmed case. Still manufactured by Remington.

.45 ACP Adopted by the U.S. Army in 1911. It's used in .45 caliber automatic pistols. Has a straight, rimless case.

.45 Colt Straight, rimmed case. In use for a hundred years. Preceded the .45 ACP as the U.S. Army service cartridge. Used in revolvers.

Shotgun Gauges

If a pound of lead was split into ten equal lumps, and then a second pound of lead was split into twenty equal lumps, which lumps would be larger? The ten-to-the-pound lumps, of course.

That is how shotgun gauges are determined. The lumps are formed into balls, and the diameter corresponding to ten-to-the-pound balls is 10 gauge. The smaller diameter corresponding to 20-to-the pound lead balls is 20 gauge.

The ordinary gauges are 10, 12, 16, and 20. Ten gauge shotguns are used for geese. Twelve gauge guns are the most popular size and best for general purpose. Sixteens have faded in popularity — hunters seeking more power have gone to the twelve; those seeking light weight have gone to the twenty.

There is also a .410 gauge. The .410 designation is really the caliber or bore diameter in inches and is not based on the same gauging system as other shotguns.

The .410 is smaller in diameter than any other shotgun. Its capacity to hold both powder and shot is less than other gauges and it compares poorly in range and stopping power. In those places where deer hunting is restricted to the use of shotguns loaded with slugs, the .410 is not allowed. It lacks adequate stopping power.

Choke

The purpose of a shotgun is to throw a wide pattern of shot. The wider the spread, the more likely you are to get some shot on the target. Human nature being what it is, no sooner did someone invent a way to throw a wide pattern of shot than someone else invented a way to restrict the pattern.

The method devised to concentrate the shot pattern was to constrict the muzzle end of the barrel — like a blunderbuss in reverse. The effect is similar to a nozzle on a garden hose.

There are several degrees of choke or constriction. They are all rated by the percent of pellets in the load they will place in a 30 inch diameter circle at 40 yards. A shotgun with no muzzle constriction at all is called cylinder bore. It is simply a straight tube.

Choke	Rating
Cylinder Bore	30-35%
Improved Cylinder	40-45%
Modified	55%
Full Choke	70-75%

Which is best depends on what you plan to use it for. A full choke is necessary for long waterfowl shots. Cylinder bore is best for deer slugs because any constriction at the muzzle will distort accuracy. For general upland hunting — rabbits, squirrels, birds — improved cylinder is best. In a pinch, they all work.

Recoil

In common language, recoil is called "kick." The same expanding gases that propel the bullet down the gun barrel also propel the barrel backward away from the bullet (thanks to the breech plug, which stops the gases from escaping).

Were you to lay the gun on a table and remote fire it, you would find that the mass and distance the gun moved backward was equal to the mass and distance the bullet moved forward. Every action has an equal and opposite reaction, according to Newton.

When the gun is held in shooting position, it is restrained from traveling backward by your shoulder. The thrust back into your shoulder is called recoil. The formula below tells how to calculate the amount of recoil energy for any combination of gun weight, bullet weight, and powder charge. This is not exactly equal to "felt recoil," however.

For example, a foregrip which offers a firm grip for the shooter can reduce felt recoil considerably. The shooter can absorb a lot of recoil with the hand that is holding the foregrip, thus saving his shoulder. The amount of energy absorbed is the same, but it *feels* like less.

One point worthy of mention: when shooting, hold the gun tight to your shoulder. Don't give it a running start before it connects. If you do, it will *really* kick you.

To figure recoil, you need to know the weight of the gun and the weight of all components which will be fired. For a rifle, this is the bullet weight and the weight of the powder charge. For a shotgun, it is the ounces of shot, the grains of powder, and the weight of the wad. Factory-made plastic wads weigh about 40 grains.

$$\text{Foot pounds of recoil energy} = \frac{(A/7000 \times B/C)^2}{64.32}$$

Where:

A = Weight in grains of fired components (bullet weight + gunpowder weight + wad weight). To convert ounces to grains, multiply the number of ounces by 437.5.

B = Muzzle velocity in feet per second.

C = Weight of the gun in pounds.

The following table is based on this formula.

Caliber	Gun Wt.	Bullet Wt.*	Recoil
.222	7	50	4
.30-30	6.5	150	12
.243	6.5	100	13
.44 Magnum	5.75	240	17
.30-06	8	180	20
12 Gauge	6.5	Various	26-28
.375 H & H	9	300	44
.458 Win.	9.5	500	65

*Bullet weight in grains with standard factory powder charge.

Stopping Power

In the days of black powder, greater stopping power was achieved by increasing the size of the chunk of lead thrown at the target. Today, stopping power is achieved by propelling a small-sized piece of lead at very high velocity — sometimes more than twice the speed of sound at point of impact.

In the old, low velocity days, the following factors were recognized as having the biggest impact on stopping power: bullet diameter, velocity, and bullet weight. Bullet diameter was a key factor. That is why a .44 Magnum (which is really a pistol cartridge) has noticeably more stopping power on deer than does a .30-30.

Guns which shoot large diameter, heavy bullets have poor trajectory characteristics. If you beef up the charge of gunpowder to extend the range, you increase the recoil. For this reason, higher speed cartridges were developed. It was observed that ultra-high speed bullets (striking the target at Mach 2) had much more stopping power than expected. The wound caused by such a bullet affected an area 30-40 times the diameter of the bullet itself.

The only formulas I have been able to find on stopping power relate to older, low velocity loads. Foot pounds of energy don't tell the whole story either. (Energy = half the mass x the square of the velocity.) Comparing only foot lbs. of energy would lead one to conclude that the .35 Remington, for example, has less stopping power than a .30-30 — and this conclusion does not square with observed fact.

I have combined information from several sources into the following formula. Velocities and foot pounds of energy can be obtained in manufacturer's brochures handed out in gun stores. The formula assigns an arbitrary value of 100 to both the .243 Winchester at 275 yards and to the .30-30 at 175 yards. Both are the lightest calibers at the extreme end of their ranges where they have reliable killing power on deer. Other calibers are ranked up and down from there.

Stopping power = .688 x the square root of (A x B x C x D)

Where:

A = Bullet weight in grains.

B = Velocity in feet per minute at point of impact.

C = Area of the bore in square inches.

D = Hydrostatic shock factor.

		Shock Factor
1700 fpm & below		1.0
1701-2000 fpm		1.25
2001-2257 fpm		1.75
2258 & above		2.0

	Yards		
	100	175	275
.22 long rifle	28		
.22 Hornet	55		
.222	74		
.223	77		
.30 Carbine	79		
.22-250	80		
.220 Swift	83		
.357 Magnum	87		
.243	109		100
.30-30	119	100	
.35 Remington	125		
.44 Magnum	150		
.270	150		
.303 British	157		
.30-06	177		
.45-70 Govt.	192		
.375 H & H	244		
.458 Win.	297		

3

THE BASIC SINGLE SHOT FROM PIPE

This chapter presents how to make a 12 gauge shotgun from 3/4-inch pipe. Chapter 4 contains details of what size pipe to use for various other gauges and calibers.

Figure 3-1 shows an exploded view of a single shot shotgun made from pipe. The parts have letter identification and are given below. The balance of this chapter tells how to make each part, how to assemble the parts, and how to test fire the finished gun. No special tools are required outside what is to be found in the ordinary home workshop — no metal lathe is required, no milling machine, and no welder.

The parts shown in Figure 3-1 are:

A. Barrel

B. Muzzle clamp (buy)

C. Lug

D. Collar

E. Breech Plug

F. Firing Pin

G. Retaining Screw (buy)

H. Hammer

I. Pivot Pin (buy)

J. Trigger

K. Stock

L. Spring (buy)

M. Screw (buy)

N. Nail (buy)

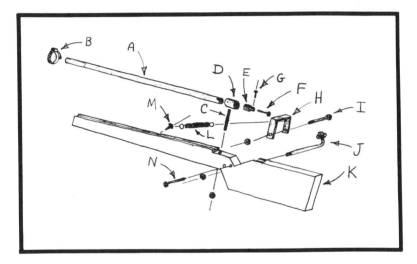

Figure 3-1

This exploded view shows each part labeled with a letter. The text refers to each part by the same letter used here.

The Barrel ("A" on the parts list)

The *inside* diameter of so-called 3/4-inch pipe is .824 inches. The *outside* diameter of a 12 gauge shotgun shell is .812 inches. The body of a 12 gauge shotgun shell will thus fit very nicely inside a 3/4-inch pipe. The rim of the shot shell is about 7/8-inches (.875) in diameter and prevents the shell from falling completely into the pipe. See Figure 3-2.

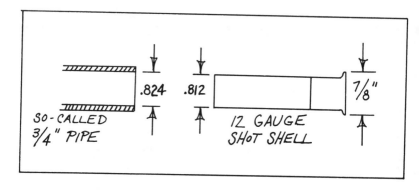

Figure 3-2

The diameter of a 12 gauge shotgun shell is .812 inches. The inside diameter of 3/4-inch pipe is .824 inches.

Hardware stores commonly sell 3/4-inch pipe in several standard, pre-cut lengths. For purposes of a 12 gauge shotgun barrel, 24 inches long is best. This is the length assumed throughout the rest of this chapter. A shorter barrel will spray shot in too wide a pattern to be effective in hunting. And, of course, the gun you make will be a cylinder bore. (See Chapter 2 for a discussion of "choke.") A barrel longer than 24 inches will not increase the pattern density to any noticeable extent but will make a far clumsier firearm to swing and point. Use only new pipe and inspect it carefully for cracks.

Figure 3-3

In the center is shown a cutaway section of pipe, showing the lip which must be ground or filed away. On top is shown an abrasive cutter in an electric drill, used to remove the lip. The piece of pipe at the bottom has had the lip removed and a 12 gauge shotgun shell now slides in smoothly.

Pipe can be cut with a hacksaw, of course, but in a hardware store it is commonly cut on a machine which works on the same principle as a tubing cutter. This machine leaves a "lip" on the end of the barrel.

The pipe you buy will probably contain such a lip. It must be reamed out. The reaming is best accomplished with a hand-held electric drill and a grinding wheel of the appropriate shape. Assortments of cheap grinding wheels come packed in plastic bags and are sold in discount department stores. See Figure 3-3.

The next topic is the threads on each end of the pipe. As it comes from the store, your precut length of pipe will have threads on each end. However, you don't need threads on the muzzle end — and the threads on the breech end are too short.

As it comes from the store, the threaded portion on each end of the pipe is 3/4-inches long. You need the breech end threaded back a full 1 1/4-inches. You can probably get the extra threads turned right at the hardware store where you buy the pipe. If an explanation seems in order, tell them your uncle is building a rack for his pickup truck and the extra threads are *his* idea — you have no idea why he wants them.

Sometimes a small hardware store will not have the machine which cuts threads. You can buy the necessary hand tools from Sears, or from a hardware or plumbing supply or industrial supply store. But go prepared to spend some money. They are not cheap. See Figure 3-4.

The extra threads *are* necessary, however. You can't do without them. The completed barrel consists of a piece of 3/4-inch pipe, 24 inches long, reamed out on each end, with 1 1/4 inches of threads on one of the two ends.

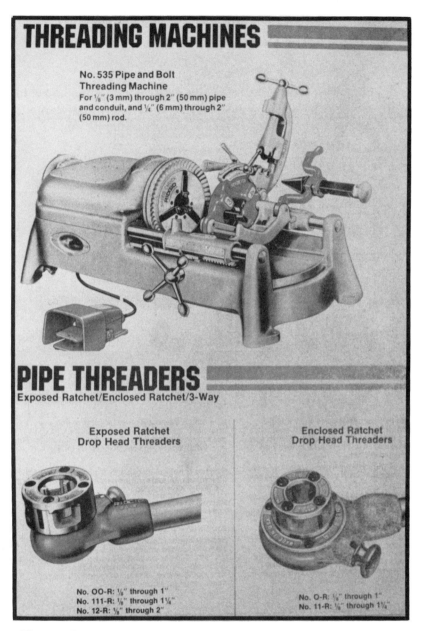

THREADING MACHINES

No. 535 Pipe and Bolt Threading Machine
For ⅛" (3 mm) through 2" (50 mm) pipe and conduit, and ¼" (6 mm) through 2" (50 mm) rod.

PIPE THREADERS
Exposed Ratchet/Enclosed Ratchet/3-Way

Exposed Ratchet Drop Head Threaders

Enclosed Ratchet Drop Head Threaders

No. OO-R: ⅛" through 1"
No. 111-R: ⅛" through 1¼"
No. 12-R: ⅛" through 2"

No. O-R: ⅛" through 1"
No. 11-R: ⅛" through 1¼"

Figure 3-4

On top is shown a threading machine such as is found in hardware stores. On the bottom is shown a hand-held pipe die and wrench such as a plumber might own. Anyone can buy them, of course, but they are not especially cheap.

The Clamp ("B" on the parts list)

This clamp is used to fasten the muzzle end of the barrel to the stock. It is simply a stainless steel radiator hose clamp 2 1/4 inches in diameter. It can be purchased at an auto parts store. See Figure 3-5.

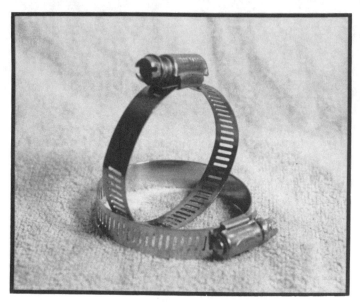

Figure 3-5
The clamp used to fasten the muzzle end of the barrel to the stock is an ordinary automotive radiator clamp.

The Lug ("C" on the parts list)

The lug screws into the collar and fastens the breech end of the barrel to the stock. It needs to be 2-3 inches long. It can be "manufactured" by sawing off the threaded portion of the bolt. Use an ordinary 5/16 inch bolt which will have 5/16 x 18NC threads. NC means National Coarse. The 18 stands for 18 threads per inch. Saw off and discard the portion not needed. See Figure 3-6.

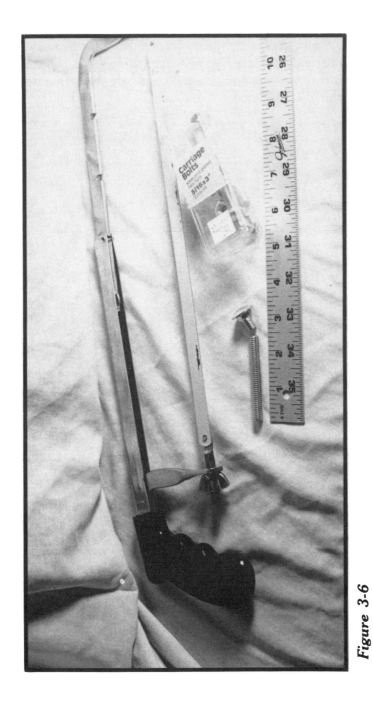

Figure 3-6
The lug is a threaded portion of 5/16-inch bolt. It needs to be two to three inches long and can be cut from an ordinary bolt.

The Collar ("D" on the parts list)

The collar is made from a steel pipe coupling. Two kinds of couplings are available (see Figure 3-7) but only one is suitable for use as a collar — the steel, threaded straight-thru type; *not* the type which is threaded on each end.

Figure 3-7

There are two types of collar available. The type on the right is more common, but the one on the left is the kind you want. It allows for solid contact between the collar and the barrel at the rear-most portion of the barrel where the explosive force to be contained is the greatest. You also want full thread engagement for the breech plug, which the style on the right does not provide.

A hole must be drilled and tapped in the collar to receive the lug (part "C"). To determine where to drill the hole, divide the length of the collar into two equal parts; then divide one of the halves into two equal parts. Using a prick punch and hammer, make a dimple on this "quarter mark." See Figure 3-8.

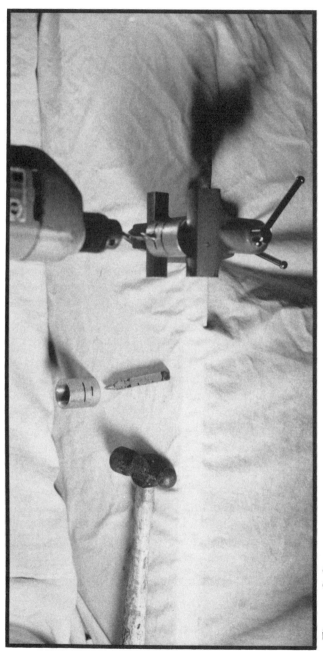

Figure 3-8
Split the length of the collar into quarters. With a prick punch, make a dimple at one of the quarter marks. Drill a 1/4-inch hole at the spot marked with the punch.

Drill a hole at this spot to be tapped for the 5/16th-inch lug (part "C"). The correct size drill bit to use is letter size F. Letter size F is .257 inches in diameter. The closest fractional-size bit is simply 1/4-inch or .250 inches in diameter. For all practical purposes, a 1/4-inch hole will work fine.

You do not drill a 5/16th-inch hole to tap for a 5/16-inch bolt. If you did, the bolt would slip in and out of the hole and the threads would not engage each other. You must drill an undersize hole so as to have some metal to fashion into threads. What size hole to drill for what size bolt is standard information in machinist's handbooks. In this case, I have done the looking up for you.

Tap threads into the hole just drilled. Use an 18-NC tap. This means (just as it did with the lug, part "C") 18 threads per inch, National Coarse. The holder which is used to grip the tap during the threading operation is called a tap wrench. The tap and tap wrench together form a T-shaped tool.

When tapping threads, begin by pushing down firmly and turning in a clockwise direction. You will feel the tool bite into the metal. Every few turns, stop and back up half a turn. This breaks up the curl of chip which is being produced so that it will fall harmlessly away. A long, curling chip can become jammed between the cutting edge of the tool and the new threads being formed, botching up the job. The lug (part "C") screws into the hole just tapped in the collar. See Figure 3-9.

It is appropriate to pause at this point and discuss the function of the lug, which is very important. "Every action has an equal and opposite reaction," according to Newton. This means that when the gunpowder in a gun "explodes," the same force which pushes the bullet down the barrel pushes the gun back into the shoulder of the shooter.

Or so it is customarily explained. A distinction is necessary. The same force which sends the bullet down the barrel pushes the *barrel* back against whatever it is that fastens the barrel to the stock. In this design, it is the 5/16-inch lug which does that.

44

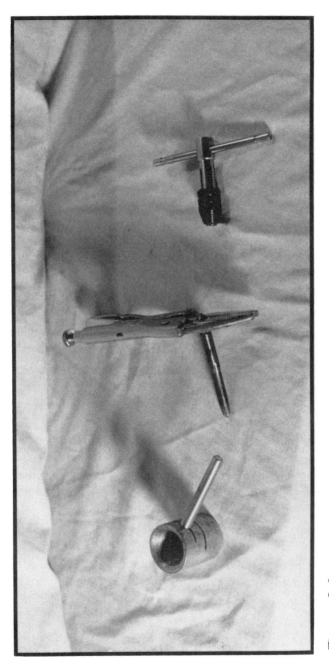

Figure 3-9

The assembled lug and collar is shown at left. The tap (used to cut threads) can be held in vise-grips as shown in the center, or clamped in a proper tap wrench, shown on the righthand side.

Many other designs exist. For example, in the *Improvised Munitions Handbook TM 31-210*, issued by the U.S. Army in 1969, string is used to fasten the barrel to the stock. Several layers of string are wrapped around the barrel-stock combination, laid on from muzzle to breech as if done by a level-wind bait casting reel, and heavily shellacked between each layer.

Personally, I much prefer the metal lug design shown here. It is mechanically more positive and, I think, safer. We talk about a gun "jumping" when it's fired. But the barrel jumping — vertically — off the stock when the gun is fired is not a concern. You do not have to fasten the barrel *down* to prevent it from jumping *up*. What you do have to worry about is the barrel traveling backward when the gun is fired and spearing the shooter in the eye. The lug design shown here looks, to me, more rugged than some factory-made firearms I have seen.

If you did not fasten the barrel to the stock, but assembled the gun otherwise complete, when you fired it the bullet would travel forward and the barrel backward. Faster than the blink of an eye, faster than any man's reflexes, the barrel would *zoom* backward about 20 feet. It's not too hard to imagine a piece of 3/4-inch pipe piercing the shooter's eye socket — the eye he used to squint down the barrel and aim — and bursting through the back of his skull.

Does that picture give you pause for thought? *Good!* With common sense and care, you can make a perfectly serviceable firearm. If you proceed slap-dash, however, you may well do more harm to yourself than you do to your intended target. The lug is a key ingredient in your own safety.

The Breech Plug ("E" on the parts list)

The breech plug is the heart of the whole mechanism. It is also the hardest to manufacture.

The breech plug performs two basic functions. First, it holds the shell in the barrel and blocks the rear end of the barrel. Second, the hole you drill in the breech plug guides the firing pin to the primer of the cartridge.

From which end of the plug should you start to drill? Figure 3-10 illustrates why it is that you should start from the round end (that is, the end which contacts the cartridge), not the square end.

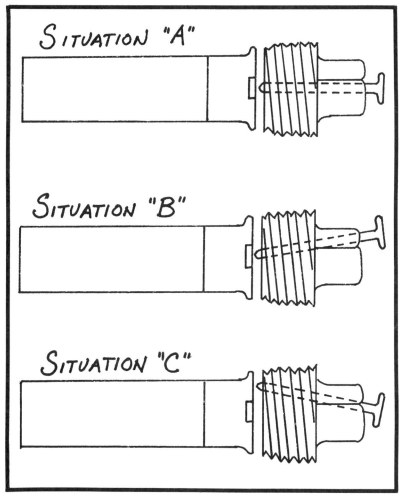

Figure 3-10

Situation "A" shows what it is you want — the firing pin hole drilled straight through the breech plug so that one end contacts the cartridge primer and the other end is centered on the rear of the breech plug. Situation "B" is less desirable, but will work. One end contacts the primer, which is essential, but the other end is not centered on the rear of the breech plug. This means that when the breech plug is screwed in, the rear of the firing pin — from the hammer's point of view — will sometimes be at 12 o'clock, sometimes at 3 o'clock, etc. Situation "C" won't work at all. Although the rear end of the firing pin is close enough to being centered to always be within reach of the hammer, the front end of the pin does not contact the cartridge primer. If you drill the firing pin hole from the side of the breech plug which contacts the cartridge, the worst situation you will end up with is Situation "B."

Situation "A" in Figure 3-10 shows what it is you *want* to end up with: a firing pin which contacts the cartridge primer on one end and which is centered on the square end as well. Being centered on the square end (that is, the rear end) of the plug ensures that the gun's hammer will make contact with the firing pin when the gun is shot.

Situation "B" shows that even if you drill crookedly, you can still produce a workable firearm *if* you start drilling on the correct end — the end which contacts the cartridge. The sketches are exaggerated, of course. I wouldn't expect anyone to really drill as crookedly as is shown.

Situation "C" shows what can happen if you start to drill from the square end of the plug and also drill crookedly. The end of the firing pin does not contact the cartridge primer and the gun will not fire.

The breech plug must be solid, to avoid excess head space. See Figure 3-11. Hollow plugs are much more common in ordinary hardware stores, but solid ones can be found. If you can't locate them in ordinary hardware stores, you will find them (for sure) in plumbing supply stores. Plumbing supply stores will sell to you, but at a high price, since they cater to professional plumbers. Do-it-yourselfers are not encouraged.

Be prepared to spend five dollars for the same item which would cost you less than one dollar if you could locate it in an ordinary hardware store.

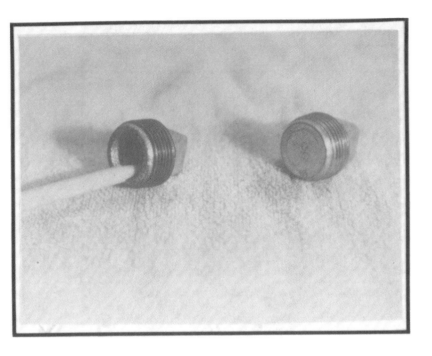

Figure 3-11

Hollow plugs, such as the one shown on the left, are easier to find than solid plugs. A solid plug is what you need, however.

The "legitimate" use of solid plugs, incidentally, is for steam lines in steam heat applications. For common water pipes, hollow plugs will suffice. If you cannot locate a solid plug, I think you should be able to fill the cavity in a hollow plug with solder or epoxy glue. I've not tried it, but I don't see why it wouldn't work.

Assuming you've obtained a solid plug, the first step in making a breech plug is to file flat the surface which will contact the rear of the cartridge. See Figure 3-12.

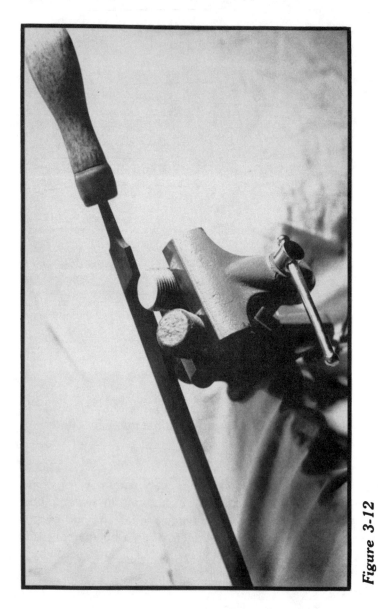

Figure 3-12

File the face of the plug flat. The plug on left is before filing. The plug on the right is after.

After that, the center of the plug must be located. To do this, a jig must be made. There are other ways to do it if you have access to machinist's tools, but this is the simplest (and most accurate) way I could devise for a home workshop. A "jig," incidentally, is a device used to locate the correct positional relationship between a piece of work and a tool.

To make this jig, start with a board about 1/2-inch thick. Cut a piece of board square, five inches on a side. If you cut the piece of wood perfectly square, then draw two lines, each connecting opposite corners, the exact center of the square will be located where the two lines cross. See Figure 3-13.

Figure 3-13

The jig which is used to locate the center of the plug is made from a piece of wood 1/2-inch thick and five inches square. Connecting the diagonally opposite corners of the square locates the center of the square.

51

After locating the center of the square, drill a 15/16-inch hole, using the square's center as the hole's center. When drilling, back up the wooden square with a piece of scrap lumber so that the drill bit won't wander when the end of the cut is reached. Backing provides the lead screw of the bit with something to bite into. See Figure 3-14.

Figure 3-14

Drill a 15/16-inch hole through the board, using the square's center as the hole's center. Back up the square with a piece of scrap lumber so the lead screw of the bit has something to bite into.

Next, screw the breech plug into the hole you have just drilled until the face of the plug is flush with the surface of the board. See Figure 3-15. Paint the face of the plug with nail polish.

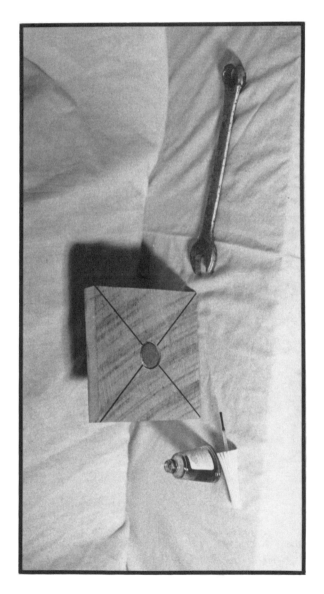

Figure 3-15

Screw the plug into the hole. Paint the face of the plug with nail polish. The nail polish provides a surface to scratch and mark the center of the plug.

After the nail polish dries, retrace the diagonal corner-to-corner lines, this time scratching them onto the painted surface. Where the lines intersect is the center of the square, the center of the hole, and the center of the plug. See Figure 3-16. Clamp the plug in a vise. Grip it by the tang — the square portion on the rear of the plug — so as not to damage the threads. Where the scratched-in lines intersect, make a dimple with a prick punch. See Figure 3-17.

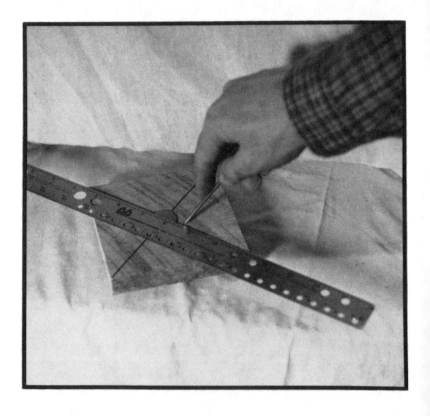

Figure 3-16

Retrace the corner-to-corner lines, scratching them into the nail polish. Where the lines cross is the center of the plug.

Figure 3-17
Dimple the center of the breech plug with a prick punch or center punch.

Drill a hole through the plug as shown in Figure 3-18. A drill press will give the best results, but a hand-held drill may be used. You might break a couple of drill bits or ruin a couple of plugs by drilling crooked holes if a hand-held drill is used, but it certainly can be done. The size hole to drill (to match the 12d nail you will use for a firing pin) is 5/32-inches.

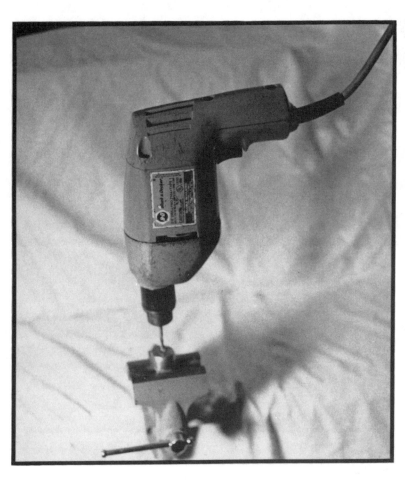

Figure 3-18

Drill a 5/32-inch hole through the plug, as shown. A drill press works better, but a hand-held drill will do the job.

The next step is to drill a hole crossways through the tang of the plug. The new hole must intersect the firing pin hole. Estimate the intersection between the two holes (see Figure 3-19) and mark with a prick punch where the new hole should be drilled.

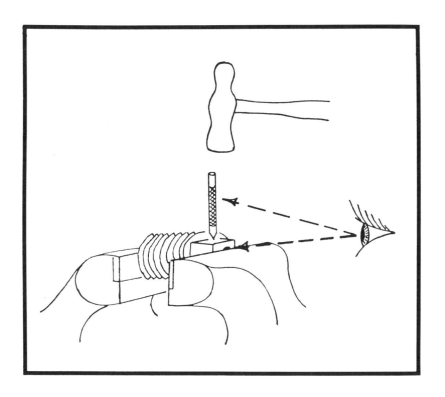

Figure 3-19
A hole is to be drilled crossways through the square tang of the plug. The new hole is to intersect the firing pin hole. Estimate, as shown, where the intersection will be and mark with a prick punch accordingly.

Drill a hole all the way through the tang of the plug. See Figure 3-20. Only half of this new hole needs threads, but the hole is drilled all the way through to provide clearance for the tap. The correct drill size to use is number size 25. A fractional 5/32-inch drill can be substituted.

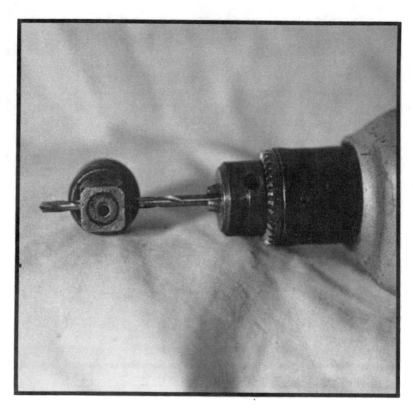

Figure 3-20

Drill the hole all the way through the tang. This provides clearance for the tap.

Tap threads into the hole just drilled. See Figure 3-21. Use a 10-24NC tap. A screw is then inserted into the tapped hole. The screw will act as a retainer for the firing pin and prevent it from falling out of the breech plug when the gun is loaded and from jumping out when the gun is fired.

Figure 3-21
Tap the threads into the tang hole, using a 10-24NC tap.

Figure 3-22
The finished breech plug with the retaining screw in place.

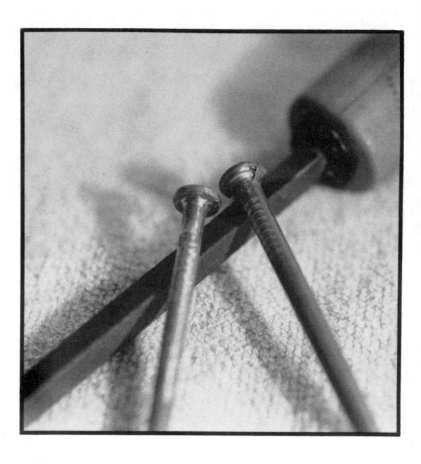

Figure 3-23

File the "head ridges" from a 12d common nail.

The Firing Pin (part "F" on the parts list)

The firing pin is made from a 12d common nail. This is an easy-to-find size and fits the 5/32-inch hole you have drilled in the breech plug. You will probably have to buy a whole pound of nails, since they are not usually sold by the piece anymore. "Box" nails are skinnier or thinner than "common" nails. You want 12d common, not 12d box. When asking in a hardware store, 12d nails are sometimes called "number 12 nails" or simply "twelves."

The first step in making the firing pin is to file off what I call the "head ridges." These are small bumps of metal left over from the manufacturing step in the factory when the nail head was formed. A small, three-cornered file works well to remove the head ridges. See Figure 3-23.

The next step is to cut the firing pin to the correct length. This is the length of the breech plug plus 1/8 of an inch. The proper length to be cut must be custom measured against each individual breech plug. See Figure 3-24. Custom measuring is necessary because plugs are not all manufactured to the same length and, even if they were, filing the end flat would destroy any uniformity which existed prior to that. Notch the nail with a file at the correct location and cut it with a hacksaw.

Figure 3-24
The firing pin must protrude through the breech plug 1/8-inch. Mark where to cut with a three-cornered file.

Next, mark the area on the nail where the flat is desired. See Figure 3-25. The flat is the portion on which the retaining screw ("G" on the parts list) will rest. The flat has to be long enough and located such that the firing pin is allowed full travel from a fully retracted position to a "fire" position — with the retaining screw in place at all times.

Figure 3-25

The flat spot on which the retaining screw lightly rests must be long enough so that the firing pin has full travel — from fully retracted to a fired position.

After estimating the flat's location, file the flat. Make it about half the thickness of the nail. Round the end of the nail which is to contact the cartridge primer and you have a completed firing pin. See Figure 3-26.

With the completion of the barrel, the collar, the breech plug, and the firing pin, you have a usable firearm of sorts. These parts alone could be assembled, held and aimed by one man, and the firing pin struck a hammer blow by a second man. English troops used such "handguns" — as distinct from

Figure 3-26
Round the end of the firing pin which contacts the primer. On the left is shown a completed firing pin. On the right is shown a completed breech plug with both firing pin and retaining screw assembled.

guns mounted on carriages — in the year 1369. One man held the "tube" or barrel while the second man applied a red-hot coal to the vent. Such guns had no stocks in this early phase of firearm development.

The Hammer ('H" on the parts list)

A good source of material for the hammer is angle iron from a hardware store. If this is what you use, buy angles (they are usually sold in pairs) that measure 4 inches on a side. To use an angle for your gun's hammer, you must straighten it out (squeezing it in a vise is a good way) and put new bends where you want them. See Figure 3-27.

Figure 3-27

The first step in making the hammer is to straighten the angle iron (corner brace) suggested in the text as "raw material."

Saw off *both* ends of the angle and discard them. See Figure 3-28. The hammer-to-be is next bent into a U-shape. As the thickness of the gunstock will be 1 1/2 inches, this is the dimension that the hammer must straddle. Therefore, let the center section of the U-shape be 1 7/8 inches wide — giving some clearance on each side of the stock — and split the

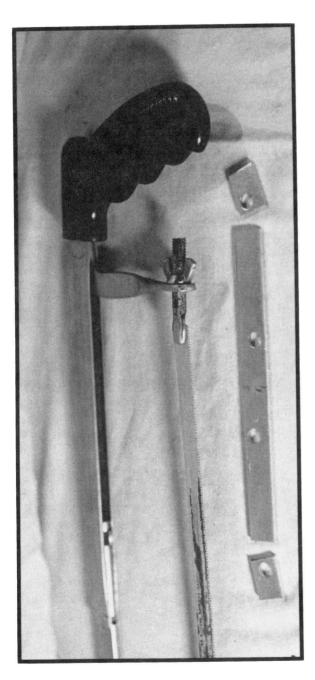

Figure 3-28

Cut off the pre-drilled hole on the straightened corner brace from each end. Don't try to use these pre-drilled holes. Without a doubt, if you do, they won't line up properly when you are finished.

remaining length between the two legs of the U. Mark the straightened angle iron accordingly and bend it in a vise. After bending, it should look like Figure 3-29.

Figure 3-29

The hammer, bent to shape. One leg will always be longer, it seems, than the other, no matter how careful you are in the marking and bending.

In spite of your best efforts to keep the legs of the U equal in length, they will never come out that way. One leg will always be shorter than the other. The ends of the legs must be drilled to accept a bolt which will act as a pivot. Drill a 1/4-inch hole close to the end of the *shortest* leg. See Figure 3-30.

It may seem silly to saw off the predrilled ends of the angle iron only to drill new holes. However, due to the fact that some angles are made with offset, non-centered holes, and the fact that the two legs cannot be bent to come out with equal length, you are better off simply sawing off the predrilled ends rather than trying to make use of them. You'll see.

Now that you have a drilled hole near the end of the shorter leg, the next thing to do is to mark the longer leg for drilling. Make a dimple using a sharpened 20d nail. Your prick punch is probably not slender enough to go through the 1/4-inch hole in the shorter leg. Everything needs to be as plumb and square as possible. The area described by the U and the 20d nail should be a rectangle with 90° corners. See Figure 3-31.

66

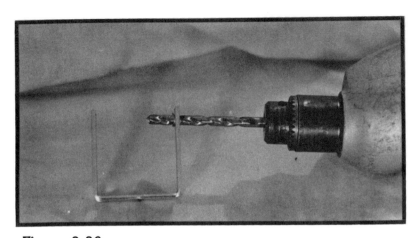

Figure 3-30
Drill a 1/4-inch hole close to the end of the shortest leg.

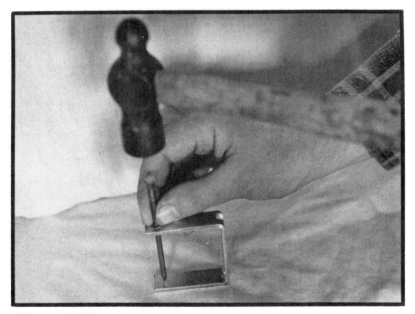

Figure 3-31
Using a sharpened 20d nail, mark the location for the second hole, as shown.

After marking, partially restraighten the angle and drill a 1/4-inch hole in the second, longer leg. See Figure 3-32.

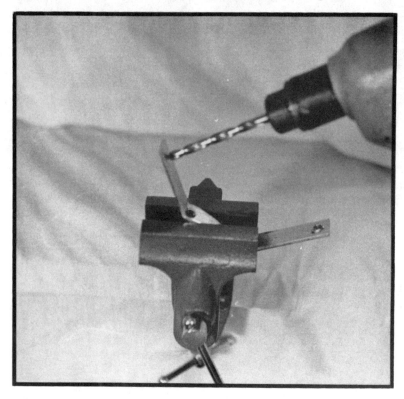

Figure 3-32

Partially restraighten the U-shape of the hammer and drill the second hole.

After drilling, rebend the U so that it once again has square corners. This is its final shape. Lastly, drill a small hole, 1/16-inch in diameter, near the edge of the hammer. See Figure 3-33. This hole will receive the end loop of the spring which pulls the hammer against the firing pin. The hammer is now complete. The pivot pin ("I" on the part list) is simply a 1/4-inch bolt, 2 1/2 inches long.

Figure 3-33

Drill a 1/16-inch hole close to the edge and close to the corner of the hammer. This hole holds the end of the spring, as shown. A 1/4-inch bolt serves as the pivot point for the hammer. Washers equalize the space between the hammer and the stock.

The Trigger ("J" on the parts list)

The trigger is optional. The gun can be fired by drawing back the hammer with the thumb and releasing it. Firing with a trigger is quicker, of course, because the hammer is already in the drawn-back position and needs only to be released. Also, holding the hammer back by hand (or, by thumb, if you will) for a long period of time, waiting for the squirrel to come back out of his hole, could be very tiring.

The trigger is made from a 1/4-inch machine bolt four inches long, plus a washer. The washer enlarges the surface which engages the hammer. See Figure 3-34. The distance from the bolt head to the bend should be about an inch.

69

Figure 3-34

The trigger is made from a machine bolt and washer. It restrains or locks back the hammer in a ready-to-fire position as shown. Installation of the trigger is an eye-balled job. Exact measurements are not given because guns made in this fashion tend not to be very exact. This figure shows how the trigger works in principle.

The Stock ("K" on the parts list)

The stock is made from a piece of 2" x 6" lumber which is 35 inches long. The dimensions are given in Figure 3-35. The dimension of 5 3/4 inches is shown on the butt portion of the stock because, in reality, that's what the width of the so-called 2" x 6" really is. The so-called 2" thickness is really 1 1/2 inches.

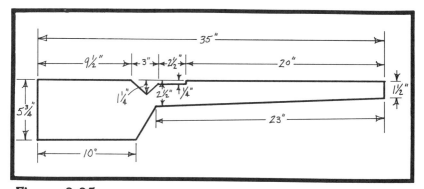

Figure 3-35

The stock is one piece, made from a sound, knot-free board. Either hardwood or softwood will work.

If possible, buy "select" quality wood. Select quality is 100% knot-free. It is used for such applications as stair treading. Softwood is okay to use. It is easy to locate, easy to work with, and strong enough for the job. Hardwood is used in factory-made guns, but is difficult to locate, difficult to work with, and will make the gun quite heavy.

Get a piece of wood which is sound — not punky, not checked or cracked. Get it as knot-free as possible. You have to use your own best judgement, then live with the consequences.

The section of stock upon which the barrel rests must be grooved for its entire length. Mark a line down the center. Then, chisel in from each side. See Figure 3-36. Do as good a job as you can, even though the results will still (in all probability) look pretty ragged. The looks will improve in the end product because the barrel will be bedded in epoxy glue which will mask the jagged carving line.

Other Parts

The remaining parts are all purchased and used as is. The spring ("L" on the parts list) must have enough tension to detonate the cartridge primer. I can't give you hard and fast rules. Such companies as W.B. Jones Spring Company of

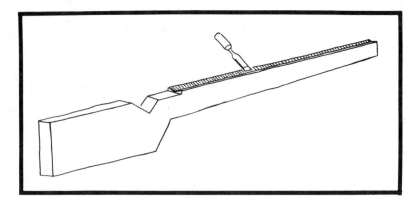

Figure 3-36

The section of stock upon which the barrel will rest must be grooved for its entire length.

Cincinnati and Century Spring Company of Los Angeles sell springs through local hardware stores. Usually the store will have a display board with a large selection of springs to look at. In a cabinet with many drawers will be springs for sale corresponding to what is on the display board. Selecting a spring is a trial-and-error proposition.

The screw ("M" on the parts list) is used to hold one end of the spring. A round head 1″ x 8″ screw is suggested.

The nail which is used as a safety ("N" on the parts list) and to hold back the hammer during loading is simply a 10d nail.

Assembly

The barrel, collar, and breech plug must be fitted with some care. If possible, when you buy your parts, take the time to screw the breech plug into the collar and the collar onto the barrel. Choose a breech plug-collar combination that screws together easily and cleanly; where the plug can be run into the collar for the full depth of its (the plug's) threads. If such a fit is not possible to find and buy ready-made, you will have to make the parts fit yourself, aided by some pipe wrenches and muscle.

The collar needs to fit onto the barrel with the same kind of snug but non-binding fit between the threads.

In assembly, first screw the collar onto the barrel. See Figure 3-37. Work it on and off until it will go on as far as necessary.

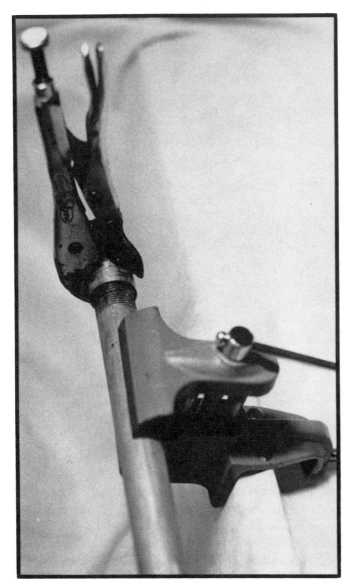

Figure 3-37
Work the collar on and off the barrel until it will screw on as far as necessary.

"As far as necessary" is far enough to have left over (not threaded onto the barrel) the thickness of a 12 gauge shell rim plus the full thread depth of the breech plug. See Figure 3-38.

Figure 3-38
The end of the barrel which receives the collar must be threaded back further than the usual 3/4-inch of threads which come from the store. The collar must be worked on the barrel until the full depth of the plug threads, plus the thickness of the shotgun shell rim, is left over. You want all of the plug threads engaged.

The breech plug must be forced into the collar in the same way. It must be run in and out until it can be screwed in full depth with the fingers. See Figure 3-39.

Figure 3-39

Work the plug into the collar until it will screw smoothly in and out to the full depth of its threads.

After the collar-barrel and the collar-breech plug combinations have been "seated" by repeated screwings and unscrewings, final assembly can begin. Screw the collar onto the barrel until it will accept a 12 gauge shell and still accept the breech plug screwed in full depth. See Figure 3-38.

It is the sheer strength of the threads which holds the breech plug in place when the gun is fired. You don't want the breech plug to pop out and hit you in the eye. You want *all* the available threads engaged, not half or two-thirds of the available threads.

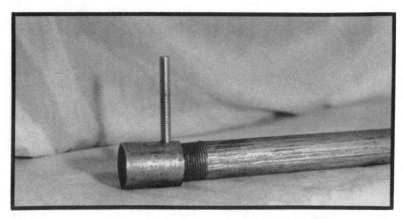

Figure 3-40

This shows the barrel, collar, and lug assembled. The lug holds the rear end of the barrel onto the stock. It is the strength of the lug which prevents the barrel from being propelled backwards when the gun is fired.

Next, screw the lug ("C" on the parts list) into the collar. See Figure 3-40. A hole must now be marked and drilled in the stock to receive the lug. Lay the barrel-collar-lug combination next to the stock and see how far to the front or rear the hole must be drilled. Mark that position on the side of the stock. See Figure 3-41. Center the hole in the side-to-side orientation. Drill a 5/16-inch hole.

Put the lug through the hole and bolt the assembly together. Next, file a shallow notch in both barrel and stock to receive the muzzle clamp ("B" on the parts list). When seated in the notches, the muzzle clamp also resists, along with the lug, the rearward thrust of the barrel when the gun is fired. See Figures 3-42 and 3-43.

Next, bed the gun barrel in epoxy glue. Dismantle the barrel from the stock; clean the barrel with sandpaper; coat with epoxy the V-shaped groove previously cut the length of the stock; reassemble — lug through hole in stock; clamp on muzzle — and let the glue dry.

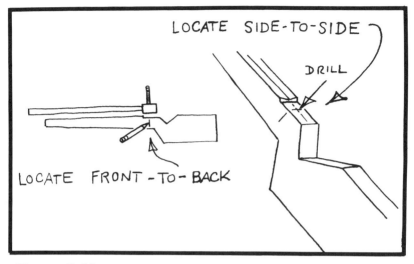

Figure 3-41

The stock must be drilled to receive the lug. Mark the front-to-back position as shown on the left. Mark the side-to-side position as shown on the right.

Figure 3-42

The muzzle clamp rests in a notch filed into the gun barrel. A corresponding notch is filed into the bottom of the stock.

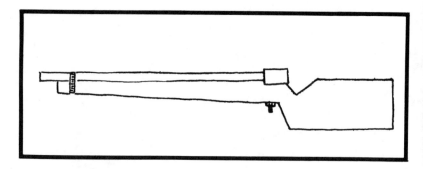

Figure 3-43

The barrel is fastened to the stock as shown: muzzle clamp on one end, lug on the other.

The next job is to install the hammer. First, insert the firing pin into the breech plug and the breech plug into the collar. The hammer must be positioned so that it will hit the firing pin. See Figure 3-10.

If the firing pin is off center, the fact that the breech plug won't be screwed in to exactly the same position each time can cause the hammer to occasionally miss the firing pin. In extreme cases, this can be corrected by modifying the hammer as shown in Figure 3-44.

Assuming that the firing pin is satisfactorily centered in the breech plug, you should position the hammer such that it (1) engages the firing pin and (2) has its pivot point directly under the firing pin. Mark the pivot point, drill a 1/4-inch hole through the stock, and install the pivot pin ("I" on the parts list). See Figure 3-45.

Three things remain to be done before the gun is ready for test firing. A spring must be installed to pull the hammer into the firing pin. A trigger must be installed (optional). And a safety must be installed.

For the spring, you have already drilled a hole in the hammer to receive one end of it. See Figure 3-33. The other

Figure 3-44

The hammer can be modified to hit an off-center firing pin.

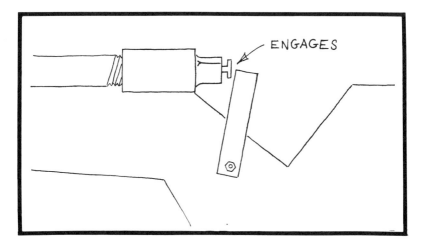

ENGAGES

Figure 3-45

The pivot point of the hammer should be directly under the firing pin.

79

end is simply secured to the stock with a screw ("M" on the parts list). Adjust the spring tension by moving the screw, trial and error. See Figure 3-46.

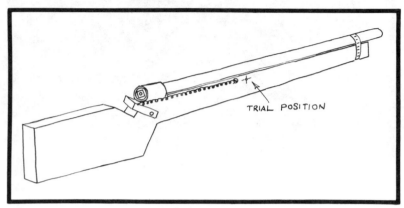

Figure 3-46

One end of the spring is held in the hole drilled in the hammer. The other end is held by a wood screw, screwed into the stock. Adjust tension by moving the wood screw. How much tension? Enough to make the gun fire.

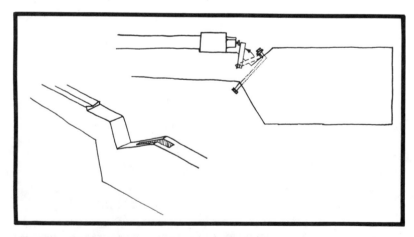

Figure 3-47

In principle, the trigger is mounted as shown. In practice, some cutting and fitting and trying will have to be done to get it to work properly.

A trigger is installed as shown in Figure 3-47. In the interest of safety, you shouldn't walk around the woods with the gun cocked and the trigger engaged. Waiting beside the woodchuck hole, of course, is a different story.

Figure 3-47 shows, in principle, how the trigger is supposed to work. In practice, your trigger will have to be custom fitted, with the stock notched or cut away as appropriate. It depends on how closely you followed the stock pattern given in Figure 3-35 and on just where and at what angle you drilled a hole in the stock to receive the trigger.

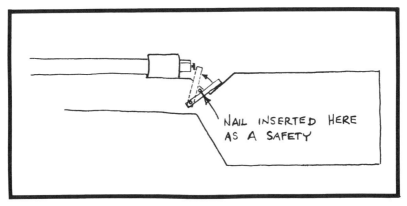

Figure 3-48
A safety — simply a nail inserted in a pre-drilled hole — can be used to hold the hammer out of the way when loading. Also, the hammer shouldn't be in hard contact with the firing pin any more than necessary.

The purpose of the safety (see Figure 3-48) is to hold the hammer out of the way while you reload. Also, when walking along with a loaded gun — not hunting, just walking, from the car to the woods, say — it is not a good idea to have the hammer in hard contact with the firing pin and the firing pin in hard contact with the cartridge primer. Should you stumble and drop the gun, for example, or set it down to cross a fence and have it fall over, it could quite possibly discharge. SURPRISE!

The completed gun is shown in Figure 3-49. It is now ready for test firing. Test firing involves remote firing of the gun. See Figure 3-50. *Do not* hand fire the gun (that is, fire the gun while holding it in your hands) until it has been remote fired at least three times. If possible, use a more powerful load in your test firing than you intend for routine use. Inspect the gun carefully after test firing. If a crack has developed, the gun is not safe to use.

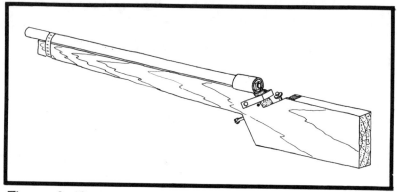

Figure 3-49
The finished firearm.

Remember, there are no guarantees! I can tell you what size hole to drill and what size tap to use. But I can't look over your shoulder and see that you do it correctly. You are the one selecting the material. You are the one putting the parts together — sloppily or carefully. You *can* make a reliable firearm. You can also make a deathtrap.

Figure 3-50

A test set-up for remote firing. The hammer is held back by the safety. The safety is released by pulling on a wire attached to it. Don't use string. String stretches, and when the nail-safety finally lets loose, it will be as though you were pulling on a rubber band. A nail, hitting you in the face at 90 MPH, is not too safe either. Note that two nails restrain the gun from being pulled over sideways when you pull on the wire. Note that a block has been nailed on to restrain the gun from traveling backwards when it is fired.

4

VARIATIONS

Chapter 3 shows how to make a 12 gauge shotgun from 3/4" pipe. Other gauges and calibers lend themselves to similar treatment. Below is a chart which gives the actual inside diameter (ID) and outside diameter (OD) of standard pipe sizes.

Back in Grandpa's day, 1/8" pipe had, approximately, a 1/8" ID, 1/4" pipe a 1/4" ID, and so forth. For reasons we need not elaborate here, the nominal pipe size (1/8", 1/4", etc.) no longer corresponds to the ID size. ID's, OD's, and wall thicknesses *have* been standardized, however, and are shown in the chart below. The OD is constant for all wall thicknesses of any given size.

Wall thickness designations of Standard, Extra-Strong, and Double Extra-Strong have been used commercially for many years. "Standard" is what you will no doubt find in the local hardware store. You will have to go to a plumbing supply house for the heavier designations. Standard is also called Schedule 40. Extra-Strong is also called Schedule 80. And Double Extra-Strong is also called Schedule 160. There is no apparent logic to much of this. It's just the way things are.

American National Standard Dimensions of Welded and Seamless Wrought Steel Pipe

Schedule 40

Size	OD	Wall Thickness	ID
1/8	.405	.068	.269
1/4	.540	.088	.364
3/8	.675	.091	.493
1/2	.840	.109	.622
3/4	1.050	.113	.824
1	1.315	.133	1.049

Schedule 80

1/8	.405	.095	.215
1/4	.540	.119	.302
3/8	.675	.126	.423
1/2	.840	.147	.546
3/4	1.050	.154	.742
1	1.315	.179	.957

Schedule 160

1/8	N/A	-	-
1/4	N/A	-	-
3/8	N/A	-	-
1/2	.840	.188	.464
3/4	1.050	.219	.612
1	1.315	.250	.815

Notes: 1. N/A means "not available."
2. All dimensions are in inches.

A word seems to be in order about the word nominal. As you can see from the above chart, nothing about so-called 1/4" pipe is actually 1/4" — not the outside diameter or the inside diameter or any characteristic. In fact, 1/4" is the "nominal" dimension only. Nominal means "for talking purposes." When you go to the hardware store and ask for 1/4" pipe, the salesman knows what you mean, and you know what you mean, and you walk out of the store with some so-called 1/4" pipe under your arm. But 1/4" is the nominal dimension only. Nothing about the pipe actually measures 1/4".

The other concept you need to understand is that of "tolerance." Nothing in the world is manufactured to the dimension called for in the blueprint (except by accident). Nothing.

Perhaps the room you're sitting in as you read this is 12 feet square. Is that the inside dimension or the outside dimension? If you measure at floor level there's the thickness of the baseboard to be considered. If you measure higher on the wall, there's the paint thickness on the wall to allow for. The paint is thicker in some spots than in others. My point is, the 12-foot dimension is only the nominal dimension. The actual dimension is 12' *plus or minus* some amount. The "plus or minus" factor is the tolerance.

When 1/4" pipe is manufactured, it is supposed to be made with an inside dimension of .364 inches. But of course it never measures that exact amount. The alloy of the metal varies slightly from lot to lot. The tooling with which the pipe is made wears from one week to the next. The speed with which different operators run the equipment varies. These factors are called "variables." And, as they change, the inside dimension of the pipe changes.

The point is this. You are using water pipe for a gun barrel. Water pipe is made with sloppier tolerances than those used for manufacturing gun barrels. The dimensions given in the above charts will be *approximately* the same as the piece of pipe you actually purchase — the operative word is "approximately."

Straight-sided shells with rims will work the easiest in homemade firearms. Below is a listing of straight-sided rimmed shells. The diameters can be compared to the pipe dimensions given above.

Rifles

Caliber	Bullet Dia	Shell OD	Rim OD
.22 Rimfire	.224	.224	.272
.375 Win.	.375	.418-.400*	.506
.444 Marlin	.429	.470-.453*	.514
.45-70 Govt.	.458	.505-.480*	.608

Pistols

Caliber	Bullet Dia	Shell OD	Rim OD
.38 Super	.355	.380	.406
.38 S&W	.357	.380	.440
.38 Special	.358	.379	.440
.357 Magnum	.358	.379	.440
.41 Magnum	.410	.434	.488
.44 Special	.433	.456	.514
.44 Magnum	.433	.456	.514
.45 Auto Rim	.454	.476-.472*	.516
.45 Colt	.454	.480	.512

Shotguns

Caliber	Bullet Dia	Shell OD	Rim OD
12 Gauge	-	.812	.875
16 Gauge	-	.750	.812
20 Gauge	-	.703	.766
.410 Bore	-	.478	.531

* indicates taper

Based on the above charts, a .22 rimfire cartridge would fit in 1/8" Schedule 40 pipe. In fact, as the diameters are the same, .22 short, .22 long, .22 long rifle, and .22 Magnum could all be fired in a gun made from such pipe.

In the case of a rimfire it is the rim, not a primer, which must be crushed to cause ignition. For the firing pin to contact the rim, it must be located off center in the breech plug. See Figure 4-1. When the breech plug is screwed into the collar, the firing pin may end up in the 6 o'clock position, the 12 o'clock position, or some intermediate position. The hammer must be wide enough to contact the firing pin in whatever position it is presented.

Figure 4-1

*The firing pin must crush the **rim** of the .22 caliber rimfire cartridge to detonate it. Thus, the hole for the firing pin in the breech plug must be drilled off center, as shown.*

A .38 S&W, .38 Special, and .357 Magnum all have *bullet* diameters that will fit in 1/4" Schedule 40 pipe. However, in each case the pipe must be drilled out to accept the diameter of the shell casing. See Figure 4-2. A 25/64" fractional drill or a letter size W drill are the bit sizes to use. Drill to a depth of 1 1/4".

A .41 Magnum *bullet* will fit in 3/8" Schedule 80 pipe. The pipe must be drilled out to accept the shell casing. See Figure 4-2. Use a 7/16" bit and drill to a depth of 1 1/4".

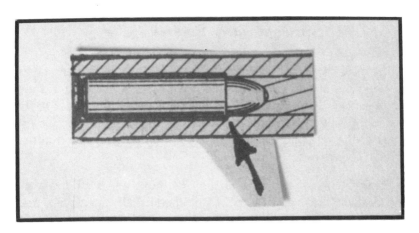

Figure 4-2

With some calibers, the bullet diameter will fit in the pipe but the shell diameter will not. The necessary clearance can be obtained by drilling out the pipe ID to the appropriate diameter and depth. The arrow in the illustration points to the potential conflict in clearance.

The .375 Winchester will fit in 3/8" Schedule 80 pipe. No drilling is necessary.

The .44 special, .44 Magnum, .45 Auto Rim, .45 Colt, and .410 bore shotgun will all fit in 3/8" Schedule 40 pipe. No drilling is required. The .444 Marlin is a sloppy fit in this size pipe and I would recommend against its use.

The .45-70 Govt. *bullet* will fit in 3/8" Schedule 40 pipe and in 1/2" Schedule 160 pipe. Both pipe sizes require drilling with a 33/64" bit to a depth of 2 1/8" for shell casing clearance.

The .35 Remington is a necked cartridge which will fit in 3/8" Schedule 80 pipe with a bit of drilling. Drill with a 15/32" bit to a depth of just over 1 9/16". The depth is fairly critical but is best determined by trial and error. It is critical because the shell is rimless and will be held in place by the shoulder instead of by a rim.

Straight-sided, Rimless Shells

Several shells have straight sides and would lend themselves to the type of firearm described in Chapter 3 except that they lack rims. "Rims" can be added, however, by the use of retaining rings. See Figure 4-3. It may be necessary to file off part of the retaining ring's "ear" to obtain clearance within the gun's collar.

The .30 M1 Carbine and the .32 Auto are both straight-sided shells that will fit into 1/4" Schedule 40 pipe. They are both rimless, but a 3/8" retaining ring can be used to solve that problem.

The *bullet* of the .380 ACP and of the 9mm Parabellum will fit in 1/4" Schedule 40 pipe. In both instances the pipe must be drilled out with a 13/32" bit to a depth of one inch to accept the shell casing. Use 3/8" retaining rings to create rims.

The .45 ACP and the .45 Winchester Magnum will fit in 3/8" Schedule 40 pipe. No drilling is required. Use 1/2" retaining rings for rims.

Retaining rings and the special pliers with which to install and remove them can be obtained at an automotive supply store. The sizes indicated here can actually be put on with the fingers and do not require special pliers (assuming you have fairly strong fingers, of course).

Necked Cartridges

By employing pipe reducers, it is possible to use necked cartridges in the type of firearm described in Chapter 3. The neck rather than the rim holds the cartridge in place. Therefore, retaining rings are not necessary even on rimless shells. Because gaps are inevitable between the cartridge and the chamber holding it, many broken shell cases and extraction problems should be expected.

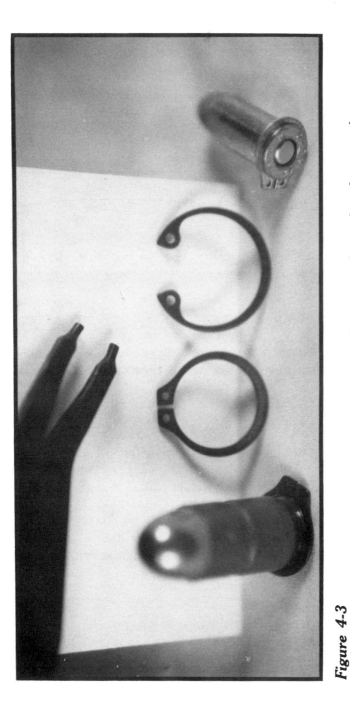

Figure 4-3
Rims can be provided by seating retaining rings in the grooves located in the rear of rimless shells. The protruding "ears" may have to be filed away to provide clearance. Two different styles of retaining rings are shown along with a special pair of pliers used to install retaining rings.

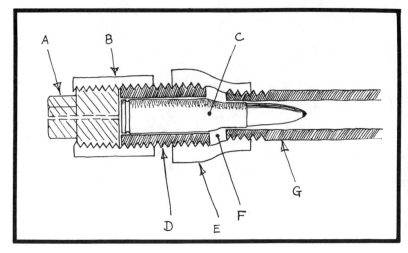

Figure 4-4

Using adapters for necked cartridges. "A" is the breech plug. "B" is the collar. "C" is the cartridge itself. "D" is the nipple which must be custom cut to length for each caliber and shell size. "E" is the reducer — the heart of the whole affair. "F" is the space, a kind of headspace, where a potential blowout of the shell casing can occur. "G" is the barrel.

When using the following directions, refer to Figure 4-4 to see the role being played by the component in question. These directions deal with diameters, not lengths. It is suggested that the nipple be 3" long to start, sawn to approximate length, and filed to finish length — being assembled and disassembled as needed for trial measurements each step of the way. In each case, the collar and plug required is the same nominal size as the large end of the reducer.

For .22 Hornet and .22K Hornet (pistol shell): use 1/8" Schedule 40 pipe for barrel; 1/4" Schedule 80 nipple; and a 1/4" to 1/8" reducer.

92

For .222, .222 Remington Magnum, .223, .221 Remington Fireball (pistol shell): use 1/8" Schedule 40 pipe for barrel; 1/4" Schedule 40 nipple drilled out with a 25/64" fractional or letter size W bit; and a 1/4" to 1/8" reducer.

For .22 PPC, 6mm PPC, .220 Swift: use 1/8" Schedule 40 pipe for barrel; 3/8" Schedule 80 nipple drilled out to 29/64"; and a 3/8" to 1/8" reducer.

For .225 Winchester: use 1/8" Schedule 40 pipe for barrel; 3/8" Schedule 80 nipple; and a 3/8" to 1/8" reducer.

For .22-250: use 1/8" Schedule 40 pipe for barrel; 3/8" Schedule 80 nipple drilled out to 15/32"; and a 3/8" to 1/8" reducer.

For .250 Savage: use 1/4" Schedule 80 pipe for barrel; 3/8" Schedule 80 nipple drilled out to 15/32"; and a 3/8" to 1/4" reducer.

For .257 Roberts, .25-06 Remington, and 6.5 x 55mm: use 1/4" Schedule 80 pipe for barrel; 3/8" Schedule 80 nipple drilled out to 31/64"; and 3/8" to 1/4" reducer.

For .270 Winchester, 7 x 57mm Mauser, 7mm Express Remington (.280 Rem.): use 1/4" Schedule 40 pipe for barrel; 3/8" Schedule 40 nipple; and a 3/8" to 1/4" reducer. (Note: the .270 Winchester is a sloppy fit and I recommend against using it unless a small powder charge is employed. The bullet can be pulled from a factory-loaded shell, for example, half the powder removed, and the bullet re-installed in the shell.)

For .284 Winchester: use 1/4" Schedule 40 pipe for barrel; 3/8" Schedule 40 nipple drilled out to 33/64"; and a 3/8" to 1/4" reducer.

For .30-30, .32 Winchester Special, and .30 Herrett (pistol shell): use 1/4" Schedule 40 pipe for barrel; 3/8" Schedule 80 nipple; and a 3/8" to 1/4" reducer.

For .300 Savage, .308 Winchester (7.62mm NATO), .30-40 Krag, .30-06, 8 x 57mm Mauser, and 8mm/06: use 1/4" Schedule 40 pipe for barrel; 3/8" Schedule 40 nipple; and a 3/8" to 1/4" reducer.

For .303 British: Use 1/4" Schedule 40 pipe for barrel; 3/8" Schedule 80 nipple drilled out to 13/32"; and a 3/8" to 1/4" reducer.

A Double Barrel

A quick follow-up shot is often nearly as valuable to a hunter as the firearm itself. The single shot shown in Chapter 3 can be converted to a double barrel with some rework.

The second barrel is mounted on top of the first barrel. The two barrels are fastened together by a short portion of bolt, threaded on both ends, connecting the two collars. To give the bolt, which I am terming a "lug," as much seating depth as possible, it is necessary to grind a flat spot on the threads of each barrel, thereby creating clearance for each end of the lug. Doing so makes for much trial assembly, disassembly, and re-assembly. The steps follow:

1. On a single shot already constructed, mark the collar with a prick punch where the connecting lug is to fasten. The second barrel is to be on top, remember.

2. Remove the collar from the single shot. Drill and tap for the connecting lug.

3. Re-assemble the collar and barrel so that the shell is seated properly.

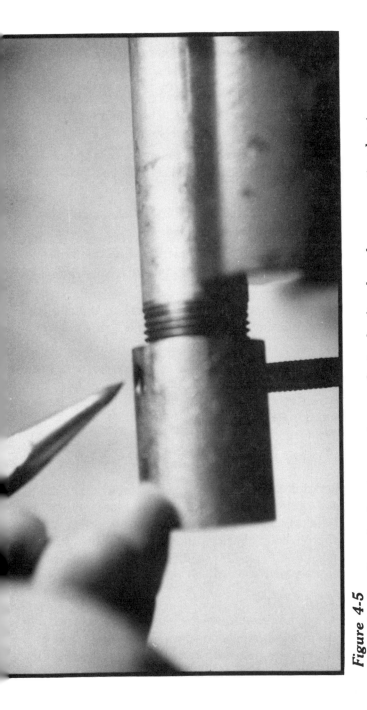

Figure 4-5
After the first breech plug is seated properly in the first barrel, a connecting lug is needed between the first and second barrels. A flat spot needs to be ground on the barrel threads for lug clearance. Where to locate the flat spot is marked with a prick punch as shown.

4. Mark the barrel threads with a prick punch where a flat spot is to be ground on the barrel threads. See Figure 4-5.

5. Again disassemble the collar and barrel and grind a flat spot on the barrel threads. Another way to create clearance for the end of the connecting lug is to use a large diameter bit and drill a shallow countersink hole. See Figure 4-6.

6. Re-assemble the barrel and collar so that the shell is seated properly *and* so that the tapped lug hole lines up with the flat spot.

7. Make a second barrel-collar-breech plug assembly in which a shell can be properly seated.

8. Mark, drill, and tap a lug hole; then disassemble and grind clearance for the end of the lug, similar to what was done on the first barrel.

9. Re-assemble the second barrel and collar, making sure the shell seats properly and that the lug hole lines up with the flat spot.

10. Measure and cut a lug. Collar thicknesses vary, so no single dimension can be given. Cut a little over-long then disassemble and re-assemble as necessary, trimming the lug length as appropriate.

11. Assemble the second barrel to the gun as shown in Figure 4-7.

12. Cut and drill a piece of hardwood as shown in Figure 4-8. The distance between the two barrel holes equals the thickness of the two collars at the rear end of the barrels. When drilling the barrel holes, back up the hardwood being drilled with a piece of scrap to prevent splitting. The small holes at the bottom are for wood screws which fasten into the end of the foregrip.

13. Insert both barrels through the hardwood and fasten into the end of the foregrip with *one* screw. That screw acts as a pivot and you can wiggle the barrels back and forth until they are properly aligned (parallel). Then, put in the second screw. See Figure 4-9.

Figure 4-6
A shallow hole (not penetrating the sidewall) can be drilled for clearance instead of grinding or filing a flat spot.

14. The next step is to make a hammer for the second barrel. Install a longer pivot pin, install both hammers on the same pin, and space them with washers as needed. Attach a spring to drive the second hammer on the opposite side of the gun from the spring which drives the first hammer. See Figure 4-10.

Figure 4-7
The two barrels are joined by screwing them together via the threaded lug and the tapped holes in the collar.

Figure 4-8

This drilled-out piece of hardwood goes on the muzzle. One barrel goes through each hole. You are making an over-and-under double barrel, not a side-by-side. The screws go into the end of the stock. The web between the two barrel holes is equal to the total combined collar thickness at the rear of the barrels, spacing the barrels apart at that end.

Figure 4-9

Getting the barrels parallel is tricky business. Insert one screw, as shown, and use it as a pivot. Twist the barrels back and forth until they are parallel, then install the second screw.

Figure 4-10
The rear end of a homemade double. The bottom hammer has been bent as required to contact the firing pin.

15. Brace the double barrel as shown in Figure 4-11 because, as described so far, only the lug connecting the collars resists the backward thrust of the top barrel when it is fired. For further reinforcement, the collars could even be welded together. The double barrel is now complete.

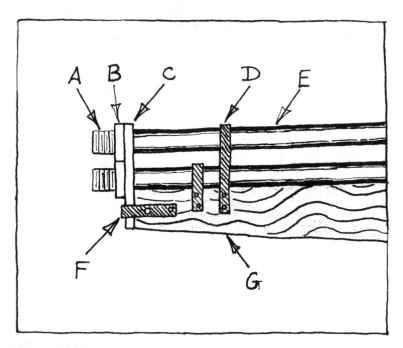

Figure 4-11

Bracing to resist the rearward thrust of the top barrel. (The lug between the collars is the only other gun part performing this function.) "A" is the threaded portion of the end of the barrel. "B" is a slice of a collar, cut with a hacksaw, here used as a retaining nut. "C" is the hardwood piece shown in Figure 4-8. "D" is a metal strap, fastening the barrel down to the stock. "E" is the top barrel itself. "F" is a metal strap which surrounds the hardwood spacer and is screwed to the stock. "G" is the wooden stock.

A Muzzle Loader

A muzzle loader has several things to recommend it. For one thing, all cartridge-type guns revolve around one thing — the cartridge. This book explains how to make gunpowder and primers and guns themselves. But it assumes you have at the very least an empty shell casing to reload.

And if you do have an empty shell casing — the heart of the whole affair — then you need a bullet the right size, a primer the right size, a gun barrel the right diameter, and a gun with the right-sized chamber to hold the shell. All this can be sidestepped with a muzzle loader. The price you pay for increased simplicity is slower repeat shots.

I recommend against building a muzzle loader from pipe with over a 1/2-inch bore. It seems to me that in large diameters the thickness of the end cap is too thin, relative to the area of the bore it must bridge, to be completely safe. Remember, test fire *everything* by remote firing before hand-held firing!

To create a muzzle loader, instead of using the collar and breech plug arrangement shown in Chapter 3, simply screw the collar further onto the barrel and cap off the breech end of the barrel as shown in Figure 4-12.

Next, drill a hole in the center of the cap, tap it for threads, and install a nipple. Directions for making a nipple are given below. The nipple is hollow and holds a percussion cap. Chapter 2 describes how a percussion gun works.

SAFETY ITEM: A factory-made muzzle loader has the nipple mounted on the *side* of the barrel. Even so, I have been hit in the eye with fragments from an exploding percussion cap and it is *not* pleasant. The design shown here puts the shooter's eye even closer to the cap. *Wear safety glasses when shooting this arm!*

Size 11 caps are the most common and the easiest to obtain. One company, incidentally, sells the Forster Tap-O-Cap for

Figure 4-12
In a muzzle loader, the function of the collar is to hold one end of the lug which fastens the barrel to the stock. A relatively long portion of the barrel is threaded. This allows the collar to be screwed completely onto the barrel and then a cap screwed on, as shown, behind the collar. The nipple mounts in a hole drilled and tapped in the pipe

about $15. With it, you can make your own size 11 caps from beer cans. They can be charged, according to the advertising, with toy gun caps, and make "a cap that is superior to anything you have ever used."

The nipple which holds the cap is threaded on one end to screw into the breech of the gun. The tap size needed for the most common nipple is 1/4 x 28NF. The correct drill size to use is a Number 3. The closest fractional drill size is 13/64".

A homemade nipple can be produced as shown in Figure 4-13 from a 40d nail. Saw off the point of the nail, tap threads onto the end, drill out the center with a 1/16" drill, cut to length, screw the nipple into the pipe cap which can then be used as a holder, and grind the end of the nipple until a number 11 cap is accepted.

To load a muzzle loading shotgun, first pour the gunpowder down the barrel. Follow this with a wad of newspaper, tamped in place with a slender dowel (ramrod). Just like tamping dynamite, don't lean over the ramrod while tamping. In case of an accident, you don't want to be speared by the ramrod.

Next pour in the shot or shot substitute (stones, BB's, ball bearings, pieces of nail, pieces of fishing sinker, etc.). The muzzle is held pointed skywards through this operation, of course, so that the shot doesn't run out on the ground. Last, insert another wad over the shot to hold it in place. Tamp this second wad in place with the ramrod.

The principle to be observed is that the wad over the powder is thick and the wad over the shot is thin — just enough to hold the shot in place. Too thick an over-shot wad will distort the pattern. How thick a wad over the powder? The spaces taken up by the wad and the shot should be equal.

Carefully remote test fire your muzzle loader before hand firing. Test the actual combination of powder, wad, and shot you plan to use in the field and then *do not exceed* what you have tested! Increasing the amount of shot, for example, will

Figure 4-13

A homemade nipple is shown screwed into a pipe cap. It has been drilled out lengthwise with a 1/16" drill bit. The top has been ground to accept a cap which simply clips on by friction. In the foreground is shown a commercial nipple (sold without restriction in sporting goods stores) and a couple of percussion caps. On the left is shown a die cutting threads on a nail and a "die stock" which is the special wrench holding the die.

106

increase the pressure which builds up inside the gun (perhaps to dangerous levels) even if no additional gunpowder is used.

An even better way of testing is to double the normal powder charge and double the normal amount of shot used. The nipple can be removed and a fuse employed to ignite the test charge. Directions for making a fuse are given below.

Oil will deaden both gunpowder and primers. Thus, if your gun is oily, fire a few caps through the empty gun to burn off any oil.

A patched ball can be fired instead of shot. This is the muzzle loading equivalent of a shotgun slug. If I remember my Revolutionary War history correctly, smoothbore muskets firing patched balls were reasonably accurate to nearly a hundred yards.

A lead ball is needed which is slightly smaller in diameter than the bore of the gun. The powder charge is first poured down the barrel, then a greased cloth patch is draped across the muzzle. The lead ball is laid on the patch and pushed into the bore until it is flush with the muzzle. The fit should be snug, not overly tight. The excess cloth is cut off flush with the muzzle. Lastly, the ball is pushed down the barrel with the ramrod and seated on the powder charge. The patch acts as a gasket and holds the gases produced by the gunpowder behind the ball.

When a patched ball is loaded in a muzzle loader, the space required in the bore of the gun is the diameter of the ball plus *twice* the thickness of the cloth patch. Ordinary patches sold for use in muzzle loaders are about .015" in thickness. Many different sizes of balls and of molds to make balls are offered for sale. Ball plus patch dimensions can be compared to the pipe dimensions given at the beginning of this chapter.

Fuses

A muzzle loader can be test fired by removing the nipple and inserting a fuse. This means that the barrel and end cap

107

assembly can be tested for sufficient strength before the gun is even assembled. An old tire can be used to hold the unstocked barrel during this procedure. See Figure 4-14.

Figure 4-14

Test firing a muzzleloader with a fuse before assembling. The fuse goes through the hole in the pipe cap normally used for the nipple.

A fuse can be made by soaking a length of string in a mixture of gunpowder and water, then letting the string thoroughly dry. Any of the gunpowder recipes given in Chapter 5 will work fine. A teaspoon of hot water, a teaspoon of gunpowder, and a length of string are all you need. Factory-made gunpowder will not dissolve in water, however.

Such a fuse works well only when totally dry. As the gunpowder recipes given in Chapter 5 tend to absorb moisture from the atmosphere, it is necessary to dry the fuse carefully and store it in a waterproof container such as a waterproof match box.

To waterproof the fuse, it can be dipped in wax. Common paraffin is brittle when cold, however, and will flake off. There is a product called DUXWAX on the market which stays reasonably flexible even when cold. Its purpose is to clean the feathers from poultry and ducks. It is sold by Cabela's, 812 13th Ave., Sidney, Nebraska 69160 and by Stromberg's, Pine River, Minnesota 56474.

Homemade fuses vary widely in how fast they burn. Play it safe. Always test a length of fuse and calculate the rate of burn, then use enough fuse to give yourself plenty of time to get to safety.

A heavy piece of yarn will make a more reliable fuse than a lightweight piece of string. Although the string will burn fine in the open air, wherever it touches another object it tends to go out. The object it touches acts as a heat sink and absorbs so much heat, in relation to what the fuse produces, that the temperature drops to below kindling point and the fuse sputters out.

A heavier fuse produces more heat and helps overcome this tendency. Only experimentation will tell you what works and what doesn't. The heat sink principle is especially worrisome at the point where the fuse enters the gun itself. If it goes out anyplace, it will probably be there.

If a fuse misfires, don't run over immediately and poke around to see what went wrong. Wait five minutes before investigating, just as you would with any misfire. One more time, be careful!

Homemade fuses are temperamental at best. If at all possible, use a factory-made fuse. The standard size is 3/32" and is waterproof. Such fuses are used by rocketry and firecracker buffs and are available from Pioneer Industries. See Chapter 6 for the address.

Wooden Guns

We tend to associate firearms with the iron age, thinking that only a metal barrel could withstand the "explosive" force of gunpowder. I have seen published directions, however, describing how to make a .22 caliber zip gun entirely from wood — barrel and all. Admittedly, the barrel was wrapped with wire for reinforcement, but it was only a 1" square piece of hardwood to start with.

I can envision an extremely primitive firearm that employs no metal whatsoever. The barrel could be made from a drilled out baseball bat. All wooden bats are seasoned hardwood, but hickory is the toughest and would be the best choice.

Instead of wrapping with wire for reinforcement, it could be wrapped with rawhide. I understand that woodchuck hide is one of the toughest leathers going. It's recommended for shoestrings, for example.

To make rawhide from woodchuck hide, first skin the chuck, then scrape all fat from the hide, remove the hair as described below, and cut into a long, continuous strip. Soak it in water overnight, wind tightly around the wooden barrel while wet, and allow to dry and shrink in place.

To remove hair from a hide, cover the hair with a 2" thick layer of wood ashes, dampen with water (the water and ashes react to produce lye), and roll the hide up tightly, hair side in. Cover it with wet cloths and leave four days, then unroll it and scrape off the hair.

A wooden barrel thus drilled out and wrapped with rawhide, loaded as a muzzle loader, firing stones for shot, and set off with a fuse, would be a firearm employing no metal at all.

Just an interesting idea. I'm not recommending it, but I don't see why it wouldn't work. Aborigines could have firearms if they wanted. Such a weapon would be immune from metal detectors. The only problem I foresee is getting the target to stand still long enough to get a shot off.

5

GUNPOWDER

When most of us think of gunpowder, and especially homemade gunpowder, we think of old-fashioned black powder. We don't realize that black powder was called "black" because there were white powders on the market.

Black powder (and white powder, too) was very corrosive to steel, whereas modern, factory-made gunpowders are not. It is for this reason that factory-made powders, produced from fairly exotic and dangerous materials, have replaced the old-time powders.

But from a performance point of view, black powder worked quite well in firearms. And so did many of its competitors or substitutes. What I'm providing in this chapter are five super-simple gunpowder recipes. No knowledge of chemistry is required but, as always, I do urge you to follow the directions, and be careful.

How Gunpowder Works

The three states of matter are solid, liquid, and gas. Many solids will turn to a gaseous state if heated sufficiently. As a gas, they take up far more room than they do as a compressed, concentrated solid.

So it is with gunpowder. It is turned into a gas by its own burning. It contains a combustible substance (fuel) and an "oxidizer" which supplies oxygen to the combustion process.

As the gunpowder turns into gas it needs room to expand and so pushes the bullet down the gun barrel ahead of itself. With old-fashioned black powder, only about half the powder turned to gas. The other half remained solid and was seen as smoke particles. With modern, "smokeless" powder, nearly all the powder turns to gas. That is why smokeless powder is more powerful than black.

The difference between a low explosive and a high explosive is the length of time it takes for the solid to turn into a gas. With a low explosive (used in guns), the burning rate is slow enough that the bullet is forced to move down the barrel ahead of the expanding gas.

On the other hand, if a rifle cartridge were accidentally loaded with a high explosive (as used in bombs), the burning gunpowder would expand *so fast* that the bullet wouldn't be able to get out of the way and the gun would explode. The high explosive powder would change from a solid to a gaseous state much too fast to be either safe or useful in firearms.

The recipes in this chapter make slow burning, low explosives and, in this sense, are safe gunpowder substitutes.

Black Powder

Black powder, brought to the attention of the Western world in the 1200's by Roger Bacon, is the granddaddy of them all. The recipe is simple, but manufacture on a home scale is not. Because black powder occupies such a high place in our history, I will give more space to it than it probably deserves.

An elaborate study in France in the 1500's established the following formula to be the best for black powder used in firearms: saltpeter, 75%; charcoal, 15.62%; sulfur, 9.38%.

As a practical matter, this recipe can be rounded to 75:15:10. These percentages are by *weight*, not volume.

The following is a description of how black powder was made commercially. It is intended to show how difficult it would be to make shooting-quality black powder on a home scale.

First, the ingredients were dampened and mixed by hand. Dampening kept down any dust which resulted from the mixing. Dust explosions happen even in grain elevators, not to mention gunpowder factories.

The ingredients were then "incorporated" by being rolled under stone wheels for three hours. The stone wheels weighed ten tons. This was called a wheel mill. Considerable heat was generated and water was added as necessary to keep the mixture moist.

The mixture was then pressed. Alternate layers of aluminum plates (aluminum is non-sparking) and gunpowder were placed in a hydraulic press at 1200 psi. The resulting press cakes were .75" thick and two feet square.

Granulating or corning came next and was the riskiest operation in black powder production. The corning mill was located at a distance, never approached while running, and remote controlled. The press cakes were crushed between rolls and graded through mechanically shaken screens. Coarse pieces were recrushed and rescreened.

The powder granules were polished by tumbling in a revolving wooden barrel. The powder was then dried, glazed (to retard soaking up humidity), and screened for size. Grading, in black powder, refers to grain size, not quality.

A test for black powder, used from ancient times, is to burn a little pile of powder on a flat, cold surface. A good powder will burn in a flash and leave no pearls of residue. A residue indicates the powder was damp at the time of the test, that the powder has been wet at some time in the past, or that the ingredients are not well incorporated.

If you try to make black powder at home, the best way to mix the ingredients is in a mortar and pestle, remembering to keep the ingredients moist. If you patiently grind away for an hour or two, then grate the powder through a piece of window screen and dry it, you will produce a poor grade of black powder. When loaded in your 12 gauge it will sound like a .22 and dent a tin can at close range.

There's a booklet on the market entitled *CIA Field Expedient Preparation of Black Powders.* In my experience, it doesn't work — it is junk. There is no way on home scale you can incorporate the ingredients properly the way a wheel mill does on commercial scale. If you really need gunpowder, my advice is not to waste your resources trying to make the black variety.

Crushed Match Heads as Gunpowder

Probably the very simplest way of making gunpowder is to crush up match heads. Safety matches will work, as will strike anywhere matches with the sensitive tip portion removed. The tip portion is too sensitive, too fast burning, too close to being high explosive, for safe use as gunpowder.

To "manufacture" gunpowder from match heads, shave or peel the combustible material off the heads of several matches. In the case of strike anywhere matches, first remove the tip portion. The tips can be saved for use in primers, as explained in Chapter 6. Be careful removing the tips from strike anywhere matches. You would not feel too clever if a tip you were removing flared up and fell into a pile of others previously removed.

After the match head material has been removed from the match stick, simply dice it up into a powder with a knife. Unlike a primary explosive which is sensitive to both friction and shock, the danger of accidental ignition with this material is very small — if you wait for safety matches to ignite from banging them together, you will wait a long time.

114

The crushed up match head material is gunpowder. It can be loaded as is into a rifle cartridge and fired. Sounds too simple to work, doesn't it?

There is less recoil or "kick" in a firearm loaded with match head powder, compared to factory loads. This means that match head powder is less powerful than factory loads and can be safely substituted on a one-to-one volume basis.

U.S. Army Manual TM 31-210 on improvised weaponry says that 58 match heads (that is, what is left of a strike anywhere match after the primary tip has been removed) will be required for a .308 Winchester cartridge.

The following chart shows how many match heads are required for other calibers. I calculated the requirement two ways: first, based on shell case volume prorated from the 58 match heads for a .308 Winchester indicated above. Second, based on a substitution rate of 1.5 match heads for each grain of smokeless powder used in typical loads. In the interest of safety the lowest of the two values appears below:

Rifle	Number of Match Heads
.22 Hornet	13
.222 Remington	26
.223 Remington	39
.243 Winchester	61
.30 M1 Carbine	16
.30-30 Winchester	45
.308 Win. (7.62mm NATO)	58
.30-06	74
.375 H&H Magnum	87
.44 Magnum	32
.45-70 Govt.	76
.458 Winchester	79
Pistols	
9mm Parabellum	8
.38 Special	15

.357 Magnum	26
.45 ACP	27
Shotguns	
12 Gauge	33
16 Gauge	30
20 Gauge	27
.410 Bore	19

A couple of incidental points. A new box of Diamond brand strike anywhere matches contains 263 matches. Also, you won't get more power by overloading the shell casing with match head powder. If you put too much of this powder into a plastic shotgun shell, you won't get more penetration — all you'll do is melt the plastic case. See Figure 5-1.

Performance Rating of Homemade Powder

If a 12 gauge load will puncture a tin can (such as a coffee can — not an aluminum beverage can) using #6 shot at 20 yards, it can be deemed a killing load for rabbits. After some experimenting, I developed a five-point rating scale. It is based on half a teaspoon of powder and one ounce of #6 shot in a 12 gauge shotgun shell.

"5" Rating: will puncture a tin can at 20 yards

"4" Rating: heavy dents, an occasional fracture

"3" Rating: significant recoil; dents

"2" Rating: a 12 gauge will sound like a .22

"1" Rating: wad will land in front of the gun or may not leave the barrel; melting of the plastic shell may occur

On this scale, crushed up match heads rate a "4."

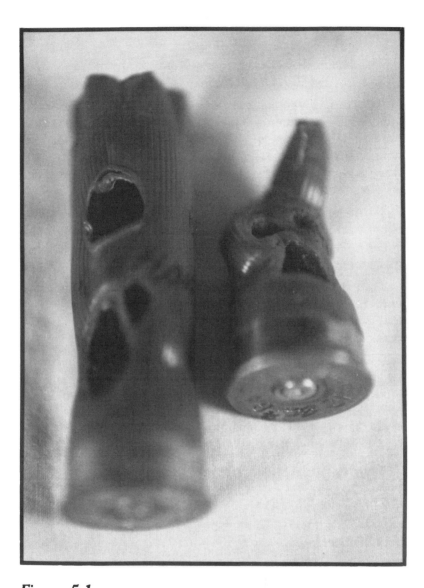

Figure 5-1

These melted shotgun shell casings resulted from attempts to get more "oomph" from weak gunpowder by adding more powder to the load. Past a certain point, you don't get more power, just more melting.

Potassium Chlorate and Sugar

This recipe yields the best gunpowder of any in this chapter. Based on the above performance scale, it rates a "5."

Its one disadvantage (aside from being corrosive, a quality which all the gunpowders in this chapter share) is that potassium chlorate is somewhat difficult to obtain. The characteristics of potassium chlorate and where to obtain it are discussed in Chapter 6. For now, it is sufficient to know that it is a white powder.

To make gunpowder by this method, simply mix together equal volumes of potassium chlorate and ordinary white table sugar. And then what? And then nothing. You have gunpowder.

Both ingredients should be dry and free of lumps. The fastest and safest mixing method I know is to pour the mixture back and forth from one saucer to another. Pour it back and forth 50 times. Use glass or plastic (non-sparking) saucers.

I have seen recipes that call for melting sugar and then adding potassium chlorate to the partially cooled but still molten sugar. I strongly recommend against it. If you've not had a frying pan of gunpowder catch fire on your kitchen stove, then you've simply not lived.

Sodium Chlorate and Sugar

This recipe is nearly identical to the preceding one. The only difference is that sodium chlorate is the oxidizer instead of potassium chlorate.

Simply mix together equal volumes of sugar and sodium chlorate. This powder rates a "5" on the performance scale. The recommended method of mixing is to pour the ingredients from one saucer to another.

Although the performance of the sodium chlorate mixture is as good as any, sodium chlorate has the drawback of being hygroscopic. This means that it picks up moisture from the atmosphere. And, as you no doubt already know, damp gunpowder doesn't shoot worth a darn.

The chemical symbol for sodium chlorate is $NaClO_3$. It is a white powder. Directions for making it at home from table salt are given later in this chapter.

To purchase commercially, sodium chlorate is more freely available than is potassium chlorate. That is, more chemical supply houses will sell it to private individuals than will do so with potassium chlorate. The two ingredients are interchangeable in most recipes.

Due to the hygroscopic problem, I would not expect a box of shells loaded with sodium chlorate powder to last more than a few days before duds began to appear. Shells can be protected somewhat with a coat of nail polish over the primer and candle wax on the over-shot wad.

Potassium Perchlorate and Sugar

The chemical symbol for potassium perchlorate is $KClO_4$. It is a white powder. Potassium perchlorate is made by heating potassium chlorate. The potassium chlorate decomposes, yielding oxygen, potassium perchlorate, and potassium chloride (common salt).

If you mix 2 parts potassium perchlorate and one part table sugar, by volume, a gunpowder results with a rating of "4." A simple mixing of powders is all that is required.

Potassium perchlorate can be purchased from Merrell Scientific. See Chapter 6 for the address.

Saltpeter and Sugar

The chemical symbol for saltpeter is KNO_3. It is a white powder. It is also called potassium nitrate, niter, nitre, nitrate of potash, and Petral stone. It is the oxidizer used in old-time black gunpowder.

If saltpeter and sugar are simply mixed together as powders, they are not incorporated together well enough to work as gunpowder. To better incorporate the ingredients, they are cooked down together like fudge candy, granulated by rubbing through a piece of window screen, then dried.

In a saucepan, combine one cup sugar, one cup saltpeter, and two cups of water. Heat with a low flame, stir with a wooden (non-sparking) spoon, and dissolve the ingredients in the water.

Bring to a low boil. Cook like fudge. If you have a candy thermometer, cook to 280° F. If you don't have a candy thermometer, cook until the consistency is semi-solid instead of liquid. Were you to cook a small batch, at the point of being "done" the "fudge" would gather around the spoon as you stirred. Try a small batch and see.

When done, pour it onto a flat surface and allow it to cool. When cool to the point it is no longer sticky, but is still moist and soft, rub pieces of it, a small gob at a time, through a piece of window screen.

Catch the granules or "worms" as they come through the screen in a shallow dish. They can be dried in the sunshine, in a food dryer, or in an oven with the temperature set as low as it will go (about 150°) and the oven door open. When dry, it is gunpowder.

If, back at the cooking stage, you cooked the "fudge" too long, hard, glassy globules will result. These can be crushed by using a hammer head as a grinding pestle on a non-sparking wooden bread board. Don't pound with the hammer; hold the head in your hand and grind with it. Sift through window screen after grinding.

Don't try and hurry the drying process by increasing the oven temperature. The top layer of the powder, closest to the heat, will start to melt. If that happens, your gunpowder will lose some of its pizzazz.

Gunpowder will last almost indefinitely if kept away from light and heat. If exposed to light or heat, it will deteriorate in only a few months. Humidity is important, too. "Keep your powder dry" is a slogan from pioneer days, but it still applies.

This powder is not as powerful as those previously listed. All of the sugar mixtures given so far require half a teaspoon of gunpowder and one ounce of shot to give best results. This recipe — saltpeter and sugar — requires 1 1/4 teaspoons of powder and an ounce of shot. Even so, it only rates a "4."

Other Oxidizers

These recipes all follow the principle of having an oxidizer mixed with a combustible. The combustible I have chosen in every case is common white table sugar. What changes from one recipe to the next is the oxidizer.

Theoretically, we should be able to substitute sodium nitrate ($NaNO_3$) for potassium nitrate in the saltpeter-sugar recipe. In practice, however, the "fudge" made with sodium nitrate is gummy and can't be granulated.

Potassium permanganate ($KMnO_4$) showed promise but produced a surprise. It is a grayish powder that turns bright purple if water is added. As a simple mixture of powders (that is, sugar and potassium permanganate) it resulted in a "2" rating. That showed promise, since many if not most oxidizers when simply mixed with sugar have a "1" rating.

To better incorporate the ingredients I tried cooking them as fudge. I put sugar in the pan. I added potassium permanganate. I added water at the sink. Before I reached the stove, however, the mixture began to spontaneously boil and smoke. It all turned to a black ash. You never know...

To save you the nuisance of experimenting, the following oxidizers rate "1" when mixed or cooked with sugar in any ratio: ammonium nitrate, strontium nitrate, barium nitrate, guanidine nitrate, barium peroxide, and potassium bichromate.

There are hundreds if not thousands of gunpowder recipes in print. Some have serious drawbacks such as becoming unstable over time. Also, although the formula may be simple, the mixing of the ingredients often isn't. Sometimes the mixing technique is not given or requires commercial machinery. In contrast, the formulas given in this chapter are simple and, most importantly, they work!

Saltpeter — Making It and Buying It

Theoretically, it is possible to extract saltpeter from both soil and from wood ashes but I have not had much success with it. Saltpeter occurs naturally in the soil in Spain, Egypt, Iran, India, Kentucky, Tennessee, and the Mississippi Valley. Soil which contains humus — well rotted animal and vegetable matter — is good candidate material, but as to whether or not saltpeter can actually be extracted, only a trial will tell.

I performed an experiment to see if the following extraction process really works. I added saltpeter to some soil which I had previously determined to be barren and was able to reclaim 11% of what I had added.

The process is to pour boiling water through the soil. The hot water dissolves the saltpeter and carries it away. The water is trapped, saved, and boiled down, thus concentrating the amount of saltpeter in the water. Next, alcohol is poured into the water and the saltpeter comes out of solution. Instead of being a clear liquid containing dissolved saltpeter, the water/alcohol mixture now has a cloudy sediment (saltpeter) in the bottom. The sediment is separated from the water by filtering through a paper filter.

If you want to try the procedure, line a kitchen strainer with a handkerchief and put the strainer over a bowl to trap the liquid which will run out. Put one cup of soil (or wood ashes) in the strainer and pour half a cup of boiling water over the soil. Pick up the corners of the handkerchief, gather it up with the soil trapped inside, and squeeze out all the liquid. It's hot, but sliding a couple of bread wrappers over your hands for insulation will help.

Repeat this process as many times as you want to obtain as much liquid as you want. Let the liquid stand overnight. In the morning it should be clear with a layer of sediment in the bottom. Siphon off the clear liquid. Discard the sediment.

Boil the liquid down to about half of its original volume. Cool to room temperature. Add an equal volume of rubbing alcohol or whiskey. A cloudy mass will form in the bottom which is the saltpeter. Pour the mixture through a paper filter. Laboratory filters, dairy filters, and coffee filters can all be used. The muddy looking sediment left on the filter is the saltpeter.

I tried various soils using this procedure and obtained nothing at all. I tried hardwood ashes and got 1 1/2 teaspoons of muddy sediment from seven cups of ashes. This turned out to be an inert substance that, even after being clarified, when made into "gunpowder" couldn't be lit with a match.

Clarification is done to the liquid before the alcohol is added. The purpose is to remove the impurities and leave the dissolved saltpeter behind in the liquid. To clarify, add egg whites to the liquid. Stir. Heat to 176° F. using a candy thermometer. The egg white cooks and traps the dirt in it. If the liquid is then strained through a handkerchief, the egg and dirt are trapped on one side; the water and dissolved saltpeter run through to the other.

Good luck. Considering the high level of frustration experienced when trying to extract saltpeter, it is fortunate that it can be purchased with very little trouble.

The "legitimate" uses of saltpeter include the brining and smoking of meat and its medicinal use as a diuretic. One brand name of saltpeter that is sold in drugstores is Purepac. It is distributed by the Purepac Pharmaceutical Co., Division of Kalipharma, Inc., Elizabeth, NJ 07207. Saltpeter can also be purchased mailorder from the chemical companies listed in Chapter 6.

People who cure meat at home use saltpeter to improve the meat's appearance. This is the same potassium nitrate which natural food advocates oppose as a harmful additive. When used as a diuretic, the dose is 1/4 teaspoon in a little water. I mention all this so that you can sound like you know what you're talking about when you go to the drugstore.

Saltpeter is also used by hospitals and institutions where it is added to cafeteria food to retard the male inmates from getting an erection. Of course, your druggist might think you a little strange if this is the reason you cite for wanting to purchase it.

Saltpeter is also sold in hardware and farm supply stores for a tree stump remover. One brand name is "Dexol." It is distributed by Dexol Industries, Torrance, CA 90501.

Making Sodium Chlorate From Table Salt

Sodium chlorate is a strong oxidizer and can be produced from common table salt via electrolysis.

Crystals of common table salt (NaCl) are comprised of sodium (Na) and chlorine (Cl). The sodium atoms carry a positive (+) charge and the chlorine atoms a negative (-) charge. Atoms which bear an electrical charge are called ions.

A battery provides direct current (DC) which flows one way. Ordinary house current is alternating current (AC) and flows first in one direction, then in the other. For the electrolysis process to work, DC current is required.

If the two poles of a battery were placed in saltwater, the sodium ions (+) would be attracted to and travel to the negative pole of the battery. The chlorine ions (-) would travel to the positive pole.

There are two kinds of conductors: those which decompose in the process of carrying electricity and those which do not. Saltwater will conduct electricity and is the kind of conductor which decomposes in the process of doing so.

As the salt decomposes into its component parts of sodium (a metal) and chlorine (a gas), the sodium reacts with the water to form sodium hydroxide (NaHO). The common name for sodium hydroxide is caustic soda or lye.

A second reaction occurs between the chlorine, previously liberated, and the sodium hydroxide, newly formed. The

124

sodium hydroxide (NaHO), the chlorine (Cl), and the water (H_2O) combine to form sodium chlorate ($NaClO_3$). The leftover chlorine bubbles out of solution at the positive terminal as a gas. The leftover hydrogen bubbles out at the negative terminal.

What I have just described contains some of my own theory about what happens. After pondering several chemistry texts, trying to figure it out, I finally found one which admitted, "At the cathode hydrogen gas, rather than metallic sodium, is produced, though just what happens is not fully understood."

In any event, hydrogen gas *is* generated and represents the one real danger to the process. Battery charging is another kind of electrolysis process yielding hydrogen gas and every garage attendant has a favorite horror story about exploding batteries. *Be warned!* Good ventilation must be provided in the electrolysis area and smoking, sparks, open flame, and arcs from electric motors eliminated.

The components of the electrolysis process are a source of DC electricity, an anode (+ terminal), a cathode (- terminal), a liquid medium (called electrolyte) through which the current travels between the anode and the cathode, and a cell or vessel to hold the electrolyte.

The cell must be of non-conductive material. On small scale a fish tank, plastic tub, or ceramic crock will do. On larger scale, a concrete box painted on the inside with epoxy paint can be constructed. The cell should be sized such that the electrolyte can circulate freely.

A battery charger can be used as a source of DC electricity. A rheostat (dimmer switch) is also needed to regulate the "amount of charge" being fed into the cell.

The electrolyte consists of 1/2 cup salt; 3 quarts of water; and 2 teaspoons of dilute sulphuric acid (that is, battery acid — it can be drawn from a car battery with an eyedropper). The sulphuric acid makes the electrolyte a better conductor of electricity.

To convert as much as possible of the salt to sodium chlorate, 1800 amp hours of electricity must pass through the solution per pound of salt. Half a cup of salt weighs 4.5 ounces and thus would require 506 amp hours. A 4-amp battery charger would have to run 126.5 hours to process half a cup of salt.

Ideally, the current pressure at the anode should be 30 amps per square foot. A 4-amp battery charger is only 13% of 30 amps and, therefore, the anode only needs to be 13% of a square foot or 19 square inches. The current pressure at the cathode should be 50 amps per square foot. A 4-amp charger should have a cathode of 11.5 square inches.

High current density improves the efficiency and the speed of the reaction but it also raises the temperature of the electrolyte. This is a troublesome area for home operation.

For best efficiency the electrolyte should be maintained at 133° F. If you are clever enough to rig up a thermostat, the battery charger can be switched off at 138° F. and back on at 127° F. (Of course, this introduces the problem of keeping track of how long the current has been flowing in the cell.) On small scale, it is probably best to sacrifice efficiency and run a cool cell. A large anode and cathode in relation to the amperage will result in cool operation as will increasing the distance between anode and cathode.

A cell on home-sized scale (See Figure 5-2 and 5-3) can be constructed as follows: make an anode of lead by melting 2½ lbs. of fishing sinkers in a 7" diameter aluminum fry pan. A propane soldering torch works well. The lead won't stick to the aluminum, so the lead "pancake" falls out easily when cool. Make a second lead disk for the cathode.

Use a two gallon crock (inside diameter = 8.5") for the cell itself. Put a layer of glass marbles on the bottom of the crock and lay a lead pancake on them. Connect the positive terminal of the battery charger to this pancake. This is the anode and should be the bottom. The chlorine gas released at the anode

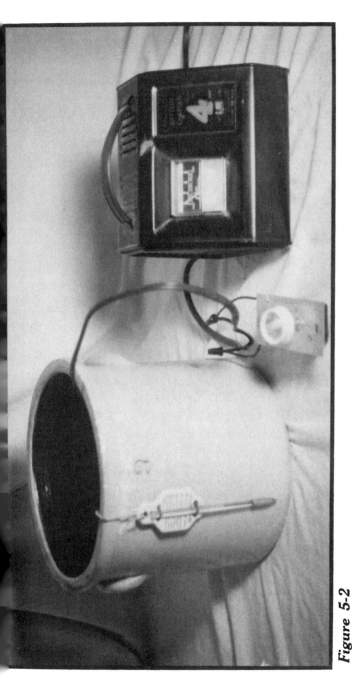

Figure 5-2
Here's what the electrolytic process looks like from the outside: a three gallon ceramic crock for a cell, a meat thermometer to monitor cell temperature, a 4-amp battery charger with dial to indicate the amount of charge, and a dimmer switch (rheostat) wired in series to regulate the input of DC electricity from the charger.

Figure 5-3
This is what is inside the electrolytic cell: a layer of marbles keeps the lower "pancake" off the bottom. A lead pancake rests on the marbles with the positive

will naturally rise to mix with the sodium hydroxide being formed at the cathode above it. An "ear" may be bent up, if necessary, to provide a good gripping place for the battery charger's clamp.

Spacers must now be provided between the anode and the cathode. These need to be one inch thick and of non-conductive material — glass, plastic, etc. Electric fence insulators, children's blocks, and medicine bottles laid on their sides offer good possibilites. Lay the cathode — the second lead pancake — on top of these spacers. Turn up an "ear," if necessary, to attach the remaining charger clamp. There should be a full inch of space between the top pancake and the clamp attached to the lower pancake. Cut away a portion of the top pancake as necessary to create this space.

Suspend a thermometer over the side of the cell to keep track of the temperature. Make up the electrolyte by putting the salt and water in a wide-mouthed gallon jar. Add the battery acid slowly. (Always add acid to water — never water to acid! This is because great heat is generated. You don't need acid spattering about.) Stir vigorously with a wooden stick (something the acid won't attack) for five minutes to *thoroughly* dissolve the salt.

Use a battery charger which has a built-in scale or dial so you can see the amount of current you are inputting to the cell. Being able to monitor current input is extremely important. Hook in a light dimmer switch to the battery charger so you can not only monitor what's going on, but control it fully, as well.

A 600 watt dimmer switch has sufficient capacity to handle a 4-amp charger and is readily available in the lighting and lamp section of department stores. Install the dimmer switch (in series) in one of the leads going from the charger to either cathode or anode. Although the scale on a 4-amp charger goes to 6 amps, don't operate your charger beyond 4 amps. You will burn out your charger if you do. Ask a man who knows.

Carefully monitor the temperature during the first day. If it gets too hot (over 138° F.) put bigger spacers between the anode and cathode. Or turn back the amps being input, using the dimmer switch.

If you don't trust your apparatus to operate safely without your vigilant presence, you can turn it off and on at will. For all practical purposes this is a non-reversible process. The sodium chlorate you are making won't revert back to saltwater just because it sits around overnight. The key is to pass 1800 amp hours of DC current through the electrolyte for each pound of salt — all at once or in a hundred small steps makes no difference.

Some of the water will be consumed in the electrolytic process. If the cathode threatens to become uncovered, add more brine solution. Otherwise, don't.

After five days (126.5 hours, to be exact) turn off the charger and filter the electrolyte through a piece of cloth. Discard any sediment in the cell. If the water is allowed to evaporate, the residue will be about 60% sodium chlorate, according to the experts — pure enough to use in gunpowder recipes.

Interestingly enough, *potassium* chlorate can be made using this same process if potassium chloride is used to start with instead of table salt (sodium chloride). Potassium chloride is sold by virtually all chemical supply houses to private individuals whereas potassium chlorate may not appear on the price list at all. Potassium chloride is also sold in grocery stores as salt substitutes for dieters.

6

PRIMERS

Ammunition is the Achilles' heel of firearms. If ammunition sales were to be restricted, guns would be rendered useless. Not so, you say, for the person who loads his own. No? What about primers? Primers are the Achilles' heel of ammunition.

Bullet material can be improvised. And gunpowder, as seen in the last chapter, can be made from scratch. But the raw materials from which primers are made pose a far greater problem. Mercury fulminate, antimony sulfide, lead peroxide, picric acid, lead azide, and nitromannite are not off-the-shelf items in your local drug store or anywhere else. These materials are difficult to obtain and dangerous to manufacture. The two methods given in this chapter represent the only primer materials I have been able to find which are both obtainable and reasonably safe.

To understand the relationship of the primer to the rest of the cartridge components, consider how you would build a fire in your fireplace. First you would light a wadded ball of newspaper. Then you would add some kindling — twigs, wood splinters, shavings. Next you would add some finely split hardwood; and lastly, the great chunks that you really wanted in the first place.

Building a fire is a progressive activity with each step of combustion based on the preceding step. Bypassing the intermediate steps — leaving out the kindling and so forth — and applying the match directly to the yuletide log is apt to have disappointing results.

The principle is the same in explosives (the progressive steps are called explosive trains) and in firearms. The combustion which takes place in a rifle cartridge is a two-step affair. The gunpowder which "explodes" and drives the bullet down the barrel is not set off by the impact of the gun's hammer. Rather, it is itself ignited by the primer. The primer contains a "primary explosive" which is sensitive enough to be set off by a spring-loaded blow from the gun's hammer. When the primer detonates, it sets off the main charge of gunpowder which is a "secondary explosive."

What considerations are involved in making your own primers? For one thing, the materials must be obtainable. It is fairly difficult for the ordinary householder to obtain a source of primary explosive.

Next, the primary explosive must be sensitive enough to do the job but relatively safe to handle. *No primary explosive is safe!* They are all, by definition, sensitive to friction, shock, heat, and spark. It is just a matter of the liability you are willing to assume when answering the question, "How risky is too risky?"

Also, the manufacturing steps must be simple enough to be accomplished with ordinary utensils and measuring capabilities. The process must be fairly speedy. The end product should not be corrosive to the firearm. And, again, the resulting primary explosive must be stable enough to be handled, stored, and loaded with reasonable safety, but sensitive enough to fire reliably. A hangfire or misfire defeats the whole purpose.

Obviously, no perfect material exists and compromises must be made. The more limited your resources — time, laboratory equipment, whatever — the more compromises must be made. Material availability is probably the single overriding factor. Corrosion considerations become a nicety, not a necessity, and can be compromised. Safety is also compromised, whether we care to admit it or not. Let's face it, the safest thing to do is to buy your ammo at the store. If you make your own, you are compromising safety. You can't have it both ways.

An acceptable limit of risk is what we seek. This chapter presents directions for making primers which, *to me*, are an acceptable compromise of all the above factors, including personal safety. I have made these primers. They work. They do have risks attached — as does everything in life. You must judge for yourself.

How a Primer Works

If you inspect a fired primer which has been removed from a center-fire rifle shell, you will notice several things: it is shaped like a miniature cup; it is dented where it was struck by the gun's firing pin; it contains a residue, left from the primary explosive, now burned away; and, inside the cup, it contains a small two or three legged piece of metal called an anvil. See Figure 6-1.

In your imagination, enlarge the primer until it is the size of a coffee cup. Imagine it to be of thin brass. The extreme tip end of strike anywhere matches is a primary explosive. Imagine match tip material had been dissolved in water, poured into the brass coffee cup like jello in a jello mold, and allowed to harden.

The coffee cup is now filled with one giant match tip. If you banged the cup on the table, would it go off? No. If you hung the cup on a string and struck the bottom with a hammer — hard; hard enough to dent the bottom — would the match tip material ignite and burn? No, it would not.

But take just one strike anywhere match tip — the very tip portion — and lay it on a blacksmith's anvil. Pound it, one blow, with a metal hammer. Will it go off? Yes. When crushed between two metal surfaces, it will explode and sound like a cap gun.

This, then, is the function of the anvil in the primer. The primary explosive must be crushed between two metal surfaces in order to detonate. The inside "back wall" of the primer cap provides one surface. The anvil provides the other.

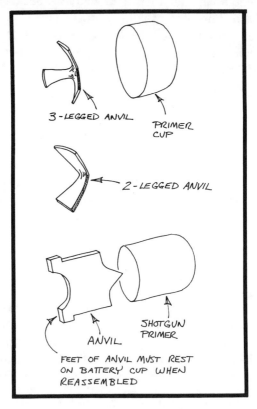

Figure 6-1

Anvils come in different shapes, but all perform the same function in the primer. They provide a surface against which the primary explosive can be crushed.

The gun's firing pin strikes the back of the primer, denting it, and crushing the trapped primary explosive between the primer wall and the anvil. The primer will not work without the anvil. When the primer is assembled in the shell casing, the shell casing itself prevents the anvil from being pushed out of the primer cup by the impact of the firing pin.

Strike Anywhere Match Tips

The modern strike anywhere match (see Figure 6-2) contains a small quantity of phosphorous trisulfide at the very tip. This is a primary explosive, sensitive to both friction and impact. When the match tip is lit, it ignites, in turn, the main body of the match head.

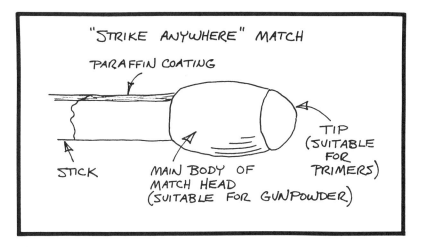

Figure 6-2
The tip portion of the common strike anywhere match is a primary explosive, and can successfully be used in primers.

The phosphorous trisulfide tip (the white part in Diamond brand matches; the light blue part in Ohio Blue Tip) is important because it will serve in primers. The rest of the strike anywhere match head (the red part of Diamond matches; the dark blue part of Ohio Blue Tip) will serve as gunpowder but is not friction sensitive or shock sensitive to a suitable degree for use as a primary explosive in primers.

Heads from *safety* matches will not work either. They are simply not sensitive enough to friction or shock for use in primers.

To reload a primer, sharpen a nail to a tapered, slender point and press out the primer as shown in Figure 6-3. The nail must be thin enough to go through the flash hole. See Figure 6-4.

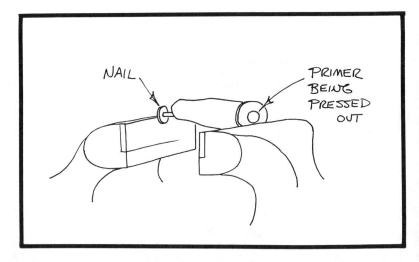

Figure 6-3

The spent primer of a rifle cartridge can be pressed out of the shell casing with a nail and a vise, as shown. The nail must be sharpened to a long, tapered point to go through the flash hole of the shell casing.

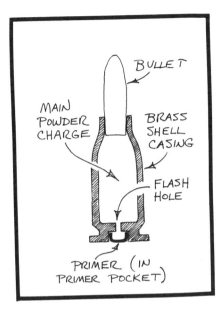

Figure 6-4

The general anatomy of a cartridge. Note the flash hole through which the nail must go to push out the old, fired primer.

Second, pry out the anvil and save it. Third, using a blunted nail, remove the dent from the primer cup as shown in Figure 6-5. You are now ready to pack the primer cup with primary explosive. To do so, you must first crush some match tips.

From three to five tips are required per primer. Primers vary in size (large rifle, small rifle, large pistol, etc.) and the amount of material deposited on the tip of the match during manufacture is not perfectly uniform from one match to another. The correct amount to use is enough to fill the primer cup level full with crushed material before tamping or compressing it.

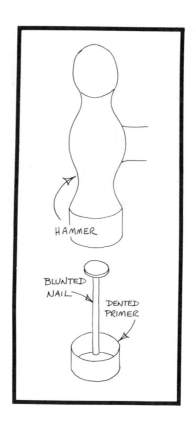

Figure 6-5

To straighten the dented primer cup, place on a hard surface and pound the dent out from the inside, as shown, using a blunted nail.

Some common sense safety procedures are in order. *Always* wear safety glasses, for example. Also, from time to time, a match tip will burst into flame while you are in the process of cutting it off the match head. When that happens, as it will, it is important that only the one tip ignites — that it doesn't fall into a pile of others previously removed.

To crush, place three to five tips on a sheet of paper on a hard surface. Make three piles. In one corner, keep the tips you have not yet started to crush. That's your raw material. In another corner, keep the crushed-up powder you have just made. That's your finished goods. The actual crushing takes place in the center, one tip at a time. That's your work-in-process. See Figure 6-6.

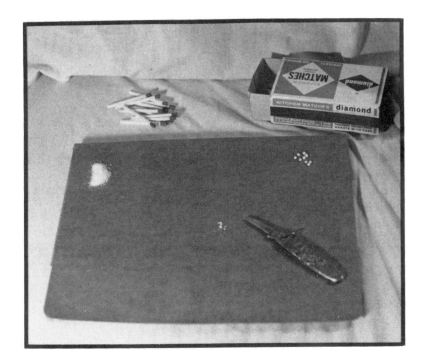

Figure 6-6

*To safely crush match tips, use three piles on a sheet of paper. One pile is for tips not yet started and one is for the final product. The third is where the tips are being crushed or chopped finely, **one at a time.***

In crushing, use the cutting edge of a knife to cut each tip in half, then in quarters, then dice to a fine powder. From time to time a match tip will ignite, especially on the first cut when the pieces are still large. If you work with three piles — raw material, finished goods, and work-in-process — the worst that can happen is that one tip ignites and burns a hole in the paper. The other way to do it, crushing several tips at once, is asking for trouble.

Use small pieces of paper for funnels, scoops, and pushers. When dealing with a very fine powder, some problems will be encountered with static electricity. Some individual particles will be repelled and you will have to chase them around to pick them up.

Fill the primer cup level full. Then use the rear end of a wooden match stick to pack down the primary explosive. See Figure 6-7. If you don't use a matchstick, use a non-metallic, non-sparking tamper as a substitute. Tamp gently at first, then push down with increasing firmness. Although the powder you are tamping is dry, it can be (and must be!) compressed to the point where none falls out when the primer cup is tipped upside down and tapped lightly.

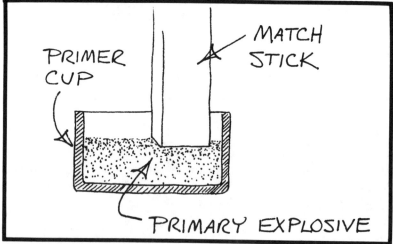

Figure 6-7

The crushed match tips are packed into the primer cup using the rear end of a wooden match stick or other non-metallic, non-sparking tamping tool.

The primary explosive in factory-made primers is coated with lacquer to seal out moisture. Although a dab of nail polish does not seem to interfere with ignition, I don't know how effective it is in sealing out moisture. Would it last ten years? There is no way of knowing, short of waiting ten years and testing it.

To reassemble the primer and cartridge, the anvil is placed not in the primer cup from where it came, but in the primer pocket in the cartridge case. See Figure 6-8. The anvil is placed in the primer pocket. Then the primer cup is placed in the primer pocket by hand, started with finger pressure, and seated with a vise as shown in Figure 6-9. You now have a primed cartridge case.

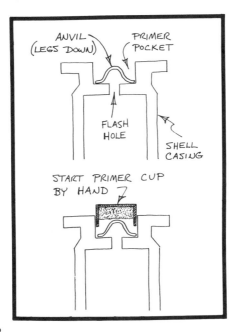

Figure 6-8

The primer and anvil are reassembled into the shell casing as shown. The anvil does not go into the primer cup, but into the shell casing "legs down." The primer cup is started into the primer pocket by hand.

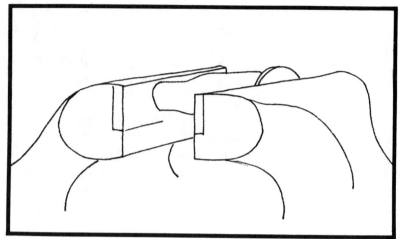

Figure 6-9

After starting the primer cup into the primer pocket by hand, it is seated by using an ordinary vise, as shown. Don't squeeze too hard and deform the end of the shell casing which is to receive the bullet.

A shotgun primer is slightly different. See Figure 6-10. The anvil must first be placed in the empty primer cup — it stays in place only by friction — and then the crushed match tips are first sprinkled in, then tamped in, around it. Reconditioning a shotgun primer is more difficult than is a rifle primer.

If you use a primer made from match tips in your gun, you must clean the gun afterwards. The residue from just one match tip is extremely corrosive. Were you not to clean the gun, within a week the bore would be coated with a thin layer of rust.

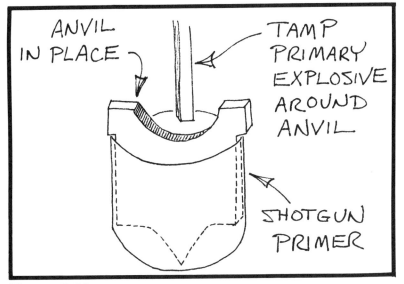

Figure 6-10

A shotgun primer is different from a rifle primer. With a shotgun, the anvil must be left in place in the primer cup (after removing the dent) and the primary explosive packed in around it. When reassembling, the "feet" of the anvil must rest on the "battery cup."

Sulfur and Potassium Chlorate

Potassium chlorate and sulfur, if mixed together, form a primary explosive suitable for use in primers. The recipe is simple, the equipment needed for measuring and blending is simple, and the materials are fairly easy to obtain — compared to the materials required for other primary explosives. On the minus side, the mixture is hazardous (which is true, of course, of any primary explosive), there are certain problems with shelf life, and it is corrosive to the metal in your gun.

First, the ingredients. Potassium chlorate (the chemical symbol is $KClO_3$) is a white powder. It dissolves in hot water.

Years ago it was used as an antiseptic for the skin and as a gargle and could be purchased off the shelf at any drugstore. Today, it is difficult to locate, although its sale is not illegal or restricted in any way. It can be purchased mail order, and firms selling it to private individuals are listed later in this chapter.

Sulfur (also spelled sulphur; the chemical symbol is S) is a pale yellow powder, famous for its rotten egg smell. It is produced in several grades, although the small companies who cater to firecracker buffs and high school chemistry students make no mention of grades. Grade A is intended for use in black powder. Grade B is intended for use in pyrotechnic compositions. Grade C is intended for use in primers. (Years ago, some commercial primers contained sulfur. None do today.)

Sulfur used to be sold in garden supply stores to make the soil acid around blueberry bushes. It has since been replaced by aluminum sulphate and can no longer be found in garden supply stores. Sulfur is still much easier to locate than is potassium chlorate, however. The kind of sulfur you want to use in primers is powdered sulfur or sulfur flour (same thing). What you do *not* want to use is "flowers of sulfur."

According to the dictionary, one of the definitions for "flower" is a finely divided powder produced by condensation or sublimation. Flowers of sulfur is also called sublimed sulfur. The point is that using flowers of sulfur in the primer mixture will yield an unstable, unpredictable explosive. Consider the following quote from **High-Low Boom!** by Philip Danisevich:

> "The combination of potassium chlorate with flowers of sulfur...form extremely sensitive mixtures, which are usually avoided, other than...in very small amounts...flowers of sulfur contain sulfuric acid...The sulfuric acid...reacts with the potassium chlorate to form highly unstable, and explosive chloric acid...(which) will eventually explode and

144

set off the remaining amount of the mixture, spontaneously. This sensitivity can be lowered by...the addition of 2% sodium bicarbonate mixed in cautiously. The sodium bicarbonate neutralizes the (sulfuric) acid present and thus eliminates the threat of chloric acid. However, mixtures of this type should not be stored for long periods of time, in large amounts."

To minimize the risk, then, use *powdered* sulfur, not flowers of sulfur. To further minimize any risk, my recipe calls for baking soda (which is the common name for sodium bicarbonate). This is to neutralize any sulfuric acid which might be present even in powdered sulfur. Risk can also be reduced by not mixing the ingredients together any further ahead of time than is necessary.

These considerations are important, unless the idea of a pocketful of shotgun shells going off spontaneously doesn't bother you. As you might expect, authoritative assessment of the risk is hard to come by. I believe you should consider all of the available information before deciding to use, or not use, a potassium chlorate-sulfur mixture.

From *The Poor Man's James Bond* by Kurt Saxon: "...potassium chlorate will also detonate spontaneously, *but not immediately*, with sulfur." (italics mine)

The book *Pyrotechny* by George W. Weingart says:
"some chemicals...fly apart...without...direct...heat. One such is potassium chlorate...This is due to the fact that its acid components, vis; chloric acid, is an unstable compound...only a slight rise in temperature is sometimes sufficient to bring about an explosion. In the presence of sulfur...which through oxidation sometimes produces minute quantities of sulfuric acid, this tendency is very strong. Consequently, compositions containing these substances must be strictly avoided."

From the *Improvised Munitions Black Book, Volume 3:* "CAUTION. Do not store the mixed explosive (potassium chlorate and sulfur) more than five days before using. KEEP THIS EXPLOSIVE DRY AT ALL TIMES."

From *The Chemistry of Powder and Explosives* by Tenney L. Davis:

> "Sulfur ought not be used in any primer composition...which contains chlorate *unless an anti-acid is present.* (italics mine) In a moist atmosphere, the sulfuric acid, which is inevitably present on the sulfur, attacks the chlorate, liberating chlorine dioxide which further attacks the sulfur, producing more sulfuric acid, and causing a self-catalyzed *souring* which results first in the primer becoming slow in its response to the trigger (hang fire) and later in its becoming inert (misfire)."

These last two quotes indicate a new problem. The mixture may sour and the cartridge primed with such a mixture may be a dud. So, according to the available experts, two extremes appear to be possible. At one extreme, your cartridges, if stored for any length of time, may go off while sitting on the shelf. At the other extreme, they might not go off at all.

Both spontaneous combustion and souring are shelf life problems. Frankly, most books don't mention either one of these problems. What they do mention is the danger involved in the act of mixing potassium chlorate and sulfur. Shelf life never enters the discussion. Still, to me at least, it seems important enough to discuss at some length.

In an effort to resolve the question, I wrote to Westech Corporation (which, unfortunately, has since gone out of business). They specialized in pyrotechnic chemicals and advertised in their catalog: "If you need any help at all, either technical, general info, or help with your order, please call us..."

The following is an excerpt from their reply: "The answer to your question is that...the longer you have them (potassium

146

chlorate and sulfur) mixed and laying around, the greater your chances for having an accident, such as spilling, dropping, etc."

Even though I had asked specifically about each, they made absolutely no mention of either spontaneous combustion or souring. From their point of view, the only danger seems to be that the longer you have the mixture sitting around, the greater the odds that you will knock it off the top of the refrigerator.

To sum up, it appears that there are risks, that we need to take steps to deal with them, but that they are not so overwhelming that we need to morbidly dwell on them night and day. One of the strongest quotes above says that mixtures containing both potassium chlorate and sulfur should be "strictly avoided." How would that same author write directions for you to store gasoline in your garage? What would he have to say about people who smoked while filling a gas tank on their lawn mower? We live with risks every day, but some common sense rules will reduce most of them to an acceptable level.

The following list contains just such rules for sulfur-potassium chlorate mixtures. These are not the only safety rules; just the ones having to do with shelf life. There are other safety rules dealing with mixing and handling. *I strongly urge you to follow them all!*

1. Use only powdered sulfur, *never* flowers of sulfur.

2. Use the purest grade chemicals available.

3. Make sure that an antacid such as baking soda is included in the mixture — 2% is recommended.

4. Keep the mixture dry.

5. Don't mix the ingredients any further ahead of time than is necessary. Five days is the maximum length of time recommended for storing either the mixture itself or primers containing the mixture.

In the case of shells which have been primed and loaded and sitting around past the five day limit, what you *should* do, ideally, is to pull the bullet and gunpowder from the shell and detonate the primer. What you *really do*, of course, is your decision. But, to be safe, the primer should be detonated.

There is one other ingredient not yet discussed: namely, ground glass. When reloading a primer, it is impossible to get the anvil back into the primer cup exactly the way it was when it came from the factory. The space between the anvil and the rear wall of the primer cup will be larger than it should be. An abrasive, such as ground glass, can be added to the recipe to give the necessary pinching effect between the particles of the mixture to cause detonation. Many old-time primer recipes call for ground glass as an ingredient.

I first attempted to obtain ground glass by rubbing a bottle on a piece of sandpaper. I realized that grit from the sandpaper came loose and fell into the glass powder. In fact, I realized that the grit from the sandpaper was just as good for the intended purpose as was ground glass — and easier to come by, too. Simply rub two pieces of sandpaper together — rough face to rough face — and capture the falling grit. A "medium" grade of paper works best.

After all the foregoing discussion and warnings, the recipe itself is quite simple. A smaller amount may be mixed if the ratio between the ingredients is maintained.

Potassium Chlorate	3	teaspoons
Sulfur	2	teaspoons
Baking Soda	1/8	teaspoon
Sandpaper Grit	1 1/2	teaspoons

All sources and authorities agree on one thing: *the act of mixing sulfur and potassium chlorate is dangerous!* Personally, I believe the real danger is that the recipe is so simple and the blending is so simple that people lose respect for it. A fancy laboratory set-up with lots of glass tubing and

148

bubbling liquids and bad smells and toxic fumes commands respect. Two simple powders, on the other hand, that you can handle separately with no risk, that you can mix at the kitchen table, and that *look* about as dangerous as cinnamon and sugar, don't command respect. Tragedy can result. This is a — BANG! — primary explosive you are dealing with.

Potassium chlorate is sometimes seen in a crystalline form rather than as a powder. The above recipe assumes the powdered form. Crystalline potassium chlorate can be crushed into a powder, but the crushing must be done between a wooden rolling pin and a wooden bread board — or similar non-sparking utensils. Be careful to clean up with a damp cloth any residue left behind. (And rinse out the cloth!) Potassium chlorate and sulfur are a primary explosive when mixed together. Don't leave smears of one to get mixed with traces of the other.

The safest way I know of to mix the above recipe is to pour the ingredients back and forth between two sheets of paper or between two saucers. See Figure 6-11. The saucers need to be of non-sparking material such as glass or plastic.

Place the sulfur, baking soda, and grit on one sheet of paper (or saucer). Then add the potassium chlorate (with any explosive mixture, the oxidizer is added last). If using paper, bend it slightly to form a trough and pour the ingredients onto a second sheet. Pour back and forth 50 times, until the ingredients are thoroughly mixed. The grit never will become uniformly mixed. Just be sure some grit is included in each primer cup that you fill. See Figure 6-12.

Use small pieces of paper for scoops, just as with match tip primers, previously discussed. Also, tamp or press the mixture into the primer cups with the rear end of a wooden match stick, just as with the match tip primers.

You will find that the sulfur-potassium chlorate mixture won't pack into the primer cups as well as the match tips do. If you push down in the center of the cup, the powder you are trying to compress squishes up around the edges. It can be very frustrating.

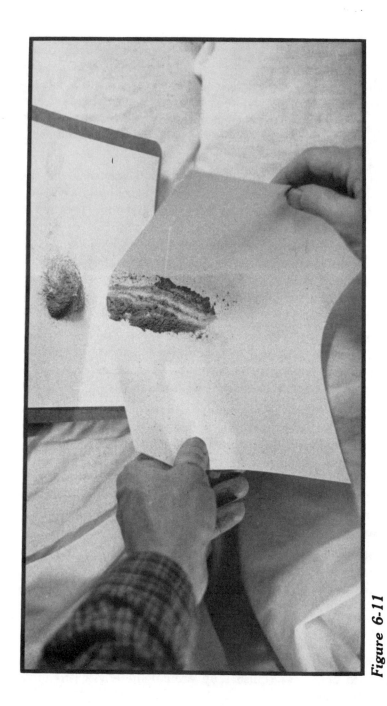

Figure 6-11

The safest way to mix sulfur and potassium chlorate is to pour

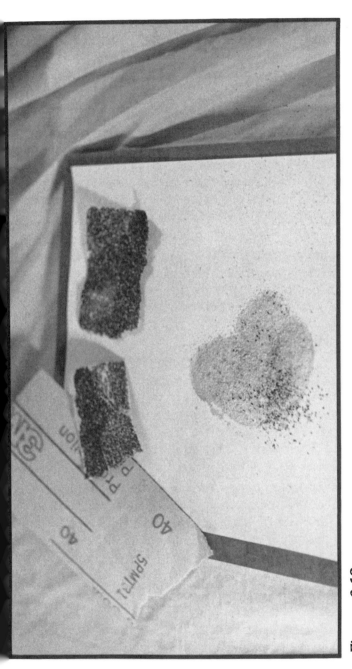

Figure 6-12

The grit, necessary to provide the pinching or crushing action which causes ignition, doesn't mix well with the other ingredients. Make sure some grit is included in each primer cup.

The solution is to pack as much in with the match stick as you can, then pack in as much more as possible with your fingertip. Lastly, paint some nail polish over the powder. Smooth it out with your fingertip. It's messy, no doubt about that. Let the nail polish dry before assembling the primer in the shell casing with the anvil.

Some safety tips: Do one primer at a time. When you are finished with it, set it down a distance away for drying. With your fingernail (*not* with a metal knife blade) scrape all residue from the outside of the primer cup while the nail polish is still damp and soft. Work on a large non-metal surface and have your materials pushed back at arm's length. Primers are normally considered throwaway items. If you attempt to recycle one, extra efforts and precautions are necessary.

To have the one primer you are working on at the moment go off accidentally might burn you, but it probably won't blind you (you might consider wearing safety glasses, of course) or burn down your house. But having one go off and set off a bunch more, or fall into a dish of primary explosive that was too near, could be disaster.

Another thing. Holding and manipulating the tiny primer cup with your fingers is difficult. However, don't use tweezers or needle-nose pliers to grip the tiny cups. Crushing the mixture between two metal surfaces is how the mixture is detonated. You would feel very foolish (wouldn't you?) if you pinched some spilled dust between the plier jaw and the outside of the primer cup and set it off.

Don't use metal containers for storage or metal utensils for handling. Use only non-sparking, non-metallic implements. I don't like the idea of glass storage containers either. The thought of glass shrapnel doesn't turn me on. I think plastic is best. Many vitamins today come packed in plastic jars which will do nicely to store your primary explosive for primers.

Assembling the finished primer in the shell casing is done exactly the same as it was for match tip primers. See Figures 6-8 and 6-9. Primers using potassium chlorate are corrosive to your gun. The same admonitions about cleaning your arm that I gave in the section on match tip primers apply here, also.

The sulfur-potassium chlorate recipe given above was developed by trial and error. It works. I have stored the mixed ingredients for over a month and found that (a) the jar did not blow up while sitting on the shelf, and (b) the primers still fired when loaded with "aged" material. Still, I must confess that I'm more comfortable when following the five-day limit.

I believe that this mixture is fairly risky to use — on par with wiring your house or racing motocross or diving in the ocean. You can kill yourself with any of these activities. But it's not *necessary* to do so. I've given some common sense rules to minimize any danger. Why not use them?

At the risk of offending some readers, an ethnic joke comes to mind which may illustrate the point. The person telling the joke pulls his hand back into his shirt sleeve until not even his fingers show. Then dangling a seemingly empty cuff, he says, "What's this?" Answer: A Polack demolitions expert. To me, if you walk around with a couple of fingers missing after fooling around with this stuff, it doesn't prove you're macho. It proves you're a Polack demolitions expert.

Mercury Fulminate

The chemical symbol for mercury fulminate is $Hg(ONC)_2$. It was the original priming material in the development of centerfire cartridges and was widely used in the 1800's. It was corrosive to the guns in which it was fired and had the additional detrimental effect on brass shell casings of making them brittle, rendering them unfit for reloading.

For the improvising reloader who wants to make his own mercury fulminate, nitric acid is required. Nitric acid is fairly dangerous to handle, difficult to obtain, and subject to many shipping restrictions if purchased mail order. It does not come in a plain brown wrapper. It comes with bells and whistles.

The recipe for making mercury fulminate is given below for those who care to try it. Personally, I don't. The idea of fooling around with nitric acid at the kitchen sink scares me, but probably no discussion of primers is complete without mercury

153

fulminate. Similar directions to those given below can be found in many old-time chemical formulary books in the public library as well as in encyclopedias.

Take mercury, 100 parts by weight; add to it nitric acid, 1000 parts by weight. The specific gravity of the nitric acid should be 1.42. Dissolve the mercury into the acid by the addition of gentle heat. It will bubble and turn green. When the solution reaches 130° F., slowly pour it through a glass funnel into alcohol, 830 parts by weight. The specific gravity of the alcohol should be .83. Red fumes will be given off first, followed half an hour later by white fumes. Five minutes after the white fumes stop, add some distilled water and filter the mixture through filter paper. White precipitate crystals are obtained which are the mercury fulminate. Wash the crystals several times with cold water. Test with litmus paper to be sure no acid remains. Dry thoroughly (at temperatures under 212° F.). The resulting product is 130% of the original weight of the mercury.

Sources of Materials

The following companies will sell potassium chlorate mailorder to private individuals. They offer free catalogs. Although I have collected a long list of chemical suppliers (I write to every lead I come across), only these two are both still in business, as well as selling potassium chlorate to private individuals.

Merrell Scientific Division
Educational Modules, Inc.
1665 Buffalo Road
Rochester, NY 14624

Pioneer Industries
Division of MNR Enterprises, Inc.
Box 36
Nashua, NH 03061

Sulfur is available from the above firms as well as:

Hagenow Laboratories
1302 Washington St.
Manitowoc, WI 54220

Hagenow offers a free catalog and has low prices. It appears to cater to the high-school-chemistry-set crowd. For the adventurous who want to experiment with mercury fulminate, all three of these firms sell mercury and nitric acid mail order to private individuals.

Recovering Potassium Chlorate From Match Heads

Potassium chlorate can be extracted from either safety matches or from strike anywhere matches that have had the tip portion removed. Match heads contain 50% potassium chlorate, in addition to glue and other ingredients.

First, soak the match heads overnight in water. Start with hot water. The next day, beat the mixture with an egg beater. The purpose is to dissolve as much potassium chlorate as possible in the water. (As long as the mixture is wet, all warnings about metal utensils — eggbeaters, for example — do not apply. When wet, the mixture is harmless. When dry, another story.)

When dissolved in water, the potassium chlorate is "in solution" in the water. A solution is different from a "mixture" in that, with a solution, the component parts cannot be mechanically separated. Filtering is an example of a mechanical process.

As a parallel example, if sugar is dissolved in coffee it is in solution. If the coffee is strained through a paper filter, the sugar comes through along with the coffee. The coffee is still sweet after filtering. Sand mixed with coffee would be an example of a mixture. If a mixture of coffee and sand were strained through a paper filter, the coffee would go through and the sand would stay behind, trapped in the filter.

The next step in the potassium chlorate recovery process, after beating with an egg beater, is to filter the match head and water slurry. A cotton handkerchief works well as a filter. Line a funnel with a handkerchief, pour the mixture through, and squeeze out all the liquid. You want the potassium chlorate, dissolved in the water, to go through and the other impurities to stay behind, trapped in the filter.

Pour the liquid into a dish and let the water evaporate. The glue from the match heads along with the other ingredients will settle out and stick to the bottom of the dish. See Figure 6-13. The top layer of this residue will be potassium chlorate crystals. They can be scraped off (using a non-metallic, non-sparking scraper), crushed, and used in the primer recipe given earlier.

Recovering Potassium Chlorate from Bleach

Potassium chlorate can be recovered from common household bleach. The Clorox label, for example, says it contains 5.25% sodium hypochlorite. That's the stuff you want.

You also need some potassium chloride. This is available from Hagenow, Merrell, and Pioneer as well as being sold as salt substitute (McCormick's, Nosalt, etc.).

Add 63 grams (that's the same as 2.2 ounces) of potassium chloride to one gallon of bleach and heat the solution to a boil. This is a simple procedure, but if you have ever spilled any bleach around the apartment, just think of the attention you will get from the neighbors if you try boiling some.

Figure 6-13

This photo shows, at the bottom, the solidified match head slurry, just as it dried onto the bottom of a glass (non-sparking) cooking dish. At the top is a pile of potassium chlorate crystals scraped off with the wooden (non-sparking) scraper. Under the dried-on crystals is a layer of sediment. This is shown, as well as a spot where the sediment has been scraped down to the glass bottom of the dish. The crystals scrape off easily, leaving the sediment behind.

Use a hydrometer to check the specific gravity. If a laboratory-type hydrometer is employed, the reading should be 1.3; if a hydrometer such as is used for automobile batteries is employed, the reading should read "full charge."

Boil the solution until the 1.3 reading is reached, then cool it in a refrigerator. Don't freeze it. The idea is that by boiling away the water you are creating a supersaturated solution. Then, by cooling it, crystals will precipitate out. Filter out the crystals and save them. They are the potassium chlorate. Repeat the process several times: boil the solution until the specific gravity reaches 1.3, then cool and filter out the crystals.

Next, mix the crystals with distilled water in the ratio of 56 grams of potassium chlorate to 100 milliliters (ml.) of water. Converted to U.S. measure, 56 grams equals 1.96 ounces and 100 ml. equals six cubic inches. The exact amount is not critical, of course, only that the *ratio* be maintained.

Heat the solution of potassium chlorate crystals and water until it boils. Allow it to cool. Then filter the crystals and save them. This step is a purification process. Dry the crystals, crush them, and you have potassium chlorate in a useable form.

Potassium chlorate can also be produced on home scale by electrolysis from potassium chloride. See Chapter 5 for details.

Softening Work Hardened Primers

If you bend a paper clip back-and-forth, back-and-forth, it will gradually become brittle and break. This phenomenon is called "work hardening" and has to do with the grain structure of the metal. All metal will work harden if bent or stretched or deformed.

Metal which has been worked hardened can be softened if it is heated past its point of recrystallization and allowed to cool slowly. This process is called annealing. The brittleness which has developed disappears, as if by magic, and the metal will again bend easily.

Primer cups, dented by the gun's firing pin and restraightened, can become work hardened and crack. Their life can be extended by heating them with a propane soldering torch until they glow a dull red and then allowing them to cool slowly to room temperature. If done, this should be performed *before* the dent made by the firing pin is straightened.

Misfires

If your gun misfires (that is, if you pull the trigger and the cartridge fails to go off) don't open the gun or take the cartridge out immediately. Wait five minutes before removing the cartridge. Be careful where you are pointing the gun while you are waiting. As soon as possible after removing the cartridge from the gun, put it to soak in a can of old motor oil. This will deaden any cartridge.

A Closing Thought

Throughout this book I have avoided editorializing. On the topic of safety, however, I feel I must comment. This is perhaps the most important chapter in this book. Certainly the information presented is the most difficult to obtain. But it is also the most dangerous chapter, from the point of view of the person doing the experimenting.

In the course of my life I have supervised a crew of men in my dad's contracting business, taught school, raised teenage children, and supervised a staff of professional engineers. And, to this day, it never ceases to amaze me how much human beings resist following directions. It seems instinctive, ingrained in human nature. "Eat your carrots. They're good for you." You know the response: "I won't." Arms folded across chest.

Well, in this case I have given you some fairly powerful information. But it is potentially dangerous to you, the user. As with life itself, there are no guarantees. You can be a Polack demolitions expert if you want — and I can't stop you. I would feel badly if someone got hurt, but I feel comfortable in my own mind that it won't happen from following the directions in this book. It will happen from *not* following the directions.

7

KITCHEN SINK RELOADING

All of the information in this book is not of much value if you don't know how to put it together. Hence, this final chapter on reloading.

A look at the reloading tools pictured in a sporting goods catalog is enough to frighten anyone. It all looks very mysterious, complicated, and intimidating. It also looks expensive.

In this chapter I will attempt to render the topic as simple as possible and show how reloading can be done with the most primitive of tools.

.22 Rimfire

Theoretically, it is possible to reload a .22 rimfire shell. The crushed rim is straightened with a punch. Using a fine artist's brush, the inside of the hollow rim is painted with a paste made from a primary explosive (see Chapter 6) and water.

More than one coat may be necessary. The primed shells are then dried (I suspect drying is the key to the success of the whole project), gunpowder is added, and a bullet inserted. It looks good on paper, but I have never seen it done successfully.

Centerfire Rifles

Figure 7-1 shows the components of a centerfire rifle cartridge. It consists of a brass shell, a primer, powder, and a bullet. How to reload primers is discussed in Chapter 6. For purposes of this discussion, we will consider the primer a non-divisible unit.

Figure 7-1

Components of a rifle cartridge. From left to right: the final, assembled cartridge, the brass shell casing, the lead bullet (in this case, a half-jacketed bullet), an empty shell on its side showing the primer pocket, a primer, and a small pile of gunpowder.

The first thing that must be done in reloading is to remove the spent primer from the fired shell. How to do this has already been shown in Chapter 6, Figure 6-3.

Next, a fresh primer is installed in the brass shell. It is started by hand with finger pressure, then seated flush as shown in Chapter 6, Figure 6-9.

Powder is then added to the primed shell casing. How much to use depends on the powder employed. For powder made from match heads, the amount to use is shown in Chapter 5.

For three other homemade mixtures (specifically potassium chlorate-sugar, sodium chlorate-sugar, and potassium perchlorate-sugar) the amount to use in a 12 gauge to achieve a killing load on rabbits is 1/2 teaspoon. The amount to use in other calibers is shown below as quantity "A."

For the saltpeter-sugar mixture, the amount to use in a 12 gauge is 1 1/4 teaspoons and the amount to use in other calibers is shown below as quantity "B." In some instances, the quantity of saltpeter-sugar mix called for might exceed the shell capacity as this is a very bulky powder. In these instances, simply load to the shell capacity.

For factory powders, load according to guidelines published in conventional reloading manuals. Manufacturers such as Speers, Hornady, and Lyman publish very complete reloading books. They are sold in gun stores.

If I were in a desperate situation and absolutely had to reload with factory powder and conventional, published reloading information was not available, I would use quantity "A," below. I would also remote fire any such load before shooting it in a hand-held firearm and I strongly urge you to do the same! See Chapter 3, Figure 3-50.

Remember:

Quantity "A" = the volume of powder to use when a 12 gauge takes 1/2 teaspoon.

Quantity "B" = the volume to use when a 12 gauge takes 1 1/4 teaspoons.

The figure given in parenthesis is the percent *by volume* of a 12 gauge load suitable for the caliber in question.

.22 Hornet (39%)

   Quantity A: 1 1/4 teaspoons split into 6 parts

   Quantity B: 1/2 teaspoon

.222 Remington (79%)

   Quantity A: 1 1/4 teaspoons split into 3 parts

   Quantity B: 1 teaspoon

.223 Remington (118%)

   Quantity A: 1/2 teaspoon + 1 1/4 teaspoons split into 12 parts

   Quantity B: 1 1/2 teaspoons

.243 Winchester (185%)

   Quantity A: 1/2 teaspoon + 1 1/4 teaspoons split into 3 parts

   Quantity B: 2 1/4 teaspoons

.30 M1 Carbine (48%)

   Quantity A: 1/4 teaspoon

   Quantity B: 1 1/4 teaspoons split into 2 parts

.30-30 Winchester (136%)

   Quantity A: 1/2 teaspoon + 1 1/4 teaspoons split into 6 parts

   Quantity B: 1 3/4 teaspoons

.308 Winchester (7.62mm NATO) (176%)

   Quantity A: 1/2 teaspoon + 1 1/4 teaspoons split into 3 parts

   Quantity B: 2 1/4 teaspoons

.30-06 (224%)

    Quantity A: 1 1/8 teaspoons

    Quantity B: 2 3/4 teaspoons

.375 H & H Magnum (264%)

    Quantity A: 1 teaspoon + 1 1/4 teaspoons split into 3 parts

    Quantity B: 3 1/4 teaspoons

.44 Magnum (97%)

    Quantity A: 1/2 teaspoon

    Quantity B: 1 1/4 teaspoons

.45-70 Govt. (230%)

    Quantity A: 1 1/8 teaspoons

    Quantity B: 3 teaspoons

.458 Winchester (239%)

    Quantity A: 1 teaspoon + 1 1/4 teaspoons split into 6 parts

    Quantity B: 3 teaspoons

9mm Parabellum (24%)

    Quantity A: 1/8 teaspoon

    Quantity B: 1 teaspoon divided into 3 parts

.38 Special (45%)

    Quantity A: 1/4 teaspoon

    Quantity B: 1 3/4 teaspoons split into 3 parts

.357 Magnum (79%)

    Quantity A: 1 1/4 teaspoons split into 3 parts

    Quantity B: 1 teaspoon

.45 ACP (82%)

    Quantity A: 1 1/4 teaspoons split into 3 parts

    Quantity B: 1 teaspoon

12 Gauge (100%)

    Quantity A: 1/2 teaspoon

    Quantity B: 1 1/4 teaspoons

16 Gauge (91%)

    Quantity A: 2 3/4 teaspoons split into 6 parts

    Quantity B: 1 1/8 teaspoons

20 Gauge (82%)

    Quantity A: 1 1/4 teaspoons split into 3 parts

    Quantity B: 1 teaspoon

.410 Bore (27%)

    Quantity A: 1/8 teaspoon

    Quantity B: 1 teaspoon split into 3 parts

I would like to stress that the above loads are derived on a calculated, theoretical basis. The 12 gauge loads have been tested and work as given. The loads for other calibers and gauges have been calculated from this and are theoretical. They are only intended to give a reasonable starting point for further testing. They should all be remote fired before hand held firing! See Chapter 3, Figure 3-50.

After remote firing, both the gun and the shell should be examined for signs of excessive pressure. If the breech plug is difficult to remove, for example, it probably indicates an overloaded situation where excessive chamber pressure has had a jamming effect on the threads. Look at the primer and compare it to an unfired shell. Excessive pressure inside the shell will have a swaging effect on the primer. With normal pressure the primer will look much the same after firing as before, except for being dented.

If you are concerned that a small amount of powder is rattling around in a relatively large case and may not be detonated by the primer, a piece of cotton fluff can be inserted after the powder and before the bullet. The fluff will hold the powder to the rear of the shell, against the primer for sure detonation. Don't overdo it. Too much wadding or packing can build up dangerous pressures.

Next comes the bullet. If you can purchase bullets, fine. If not, you can make your own. Lead is the best material, but mild steel — a nail or bolt of the right diameter and cut to the appropriate length — will work.

A bullet stays seated in a shell casing largely by friction. A sloppy fit can be improved with an aluminum foil or paper gasket. Some gentle squeezing with pliers can be done if necessary on the neck of the brass case where it grips the bullet.

From a safety and overloading point of view it is very important to use an appropriate bullet weight. Too heavy a bullet can dangerously increase the pressure within the firing chamber, even to the point of the gun exploding. The table below shows the bullet weight to use by caliber.

Caliber	Bullet Grains	Bullet Ounces	Bullets Per Lb.
.22 Horn.	40-45	.10	160
.222 Rem.	50-55	.12	133

.223 Rem.	50-55	.12	133
.243 Win.	80-90	.19	84
.30 M1	100-110	.24	67
.30-30	165	.38	42
.308 Win.	150-180	.38	42
.30-06	150-180	.38	42
.375 H&H	235-285	.59	27
.44 Mag	240	.55	29
.45-70	400	.91	17
.458 Win.	500	1.14	14
Pistols			
9mm Para	100	.23	70
.38 Spec	150	.34	47
.357 Mag	158	.36	44
.45 ACP	230	.53	31
Shotgun Shot or Slug Weight			
12 Gauge	-	1 oz.	16
16 Gauge	-	1 oz.	16
20 Gauge	-	7/8 oz.	18
.410 Bore	-	1/2 oz.	32

You can melt fishing sinkers and cast bullets in conventional bullet molds. Or, a wooden mold can be made as shown in Figure 7-2. Hardwood is best. It will scorch, but can be used several times. Lead can be melted in a pan on your kitchen stove. Solder will also work.

Cheap plastic diet scales can be used to check bullet weights, especially if the weighing is done in multiples. Your scale may not be accurate enough to weigh a quantity as small as .12 ounce, for example. But it will weigh one ounce with reasonable accuracy, and if eight equal-sized bullets weigh one ounce in total, then each bullet weighs about .12 ounces.

To summarize, the steps in loading a rifle cartridge are:

1. Remove the fired primer.

2. Install a fresh primer.

3. Load gunpowder in the shell.

4. Add a piece of cotton fluff if the volume of powder is small in relation to the shell size.

5. Install a bullet.

Figure 7-2

A homemade bullet mold. Two pieces of hardwood are clamped together and drilled to the correct diameter at the joint. Molten lead is then poured in the cavity. The bullet is removed by unclamping and pulling apart the "sandwich." The resulting slug must be trimmed and scraped to finished dimensions.

Shotgun Shells

Shotgun shells are more complicated to reload than rifle cartridges. The components are shown in Figure 7-3. They include the shell casing, the powder, the primer, the shot, and the wad.

The purpose of the wad is to trap the explosive force of the gunpowder behind the shot in such a way that the shot is expelled from the muzzle in a unit or lump. The lump of shot fans out on its way to the target, but leaves the barrel initially as a dense cluster, almost one piece.

To reload a fired shell, first remove the spent primer as shown in Figure 7-4. Next, replace it with a fresh primer. Start the new primer with finger pressure, then seat it with a vise and a length of wooden dowel as shown in Figure 7-5. Add powder as shown in Figure 7-6.

The wad is inserted next. If a conventional plastic wad is used, the mouth of the shell can be stretched by inserting and twisting a wooden rod carved to a cone shape. The wad is inserted by hand and seated with a dowel. See Figure 7-7. Be careful during these manipulations not to spill the powder which is already in the shell.

If a factory-made plastic wad is not available, a wad of paper can be used. See Figure 7-8. In a muzzle loader, a wad is used between the powder and the shot and a second wad is used after the shot to hold it in the barrel. A very similar system was used in early shotgun shells with one wad between the powder and shot and a second wad used to hold the shot in the shell. This same system could still be employed if need be.

Shot is added with a funnel. Substitutes for factory-made chilled lead shot are pieces of fishing sinker, lead collars from roofing nails, and old ball bearings.

Figure 7-3
Components of a shotgun shell. Clockwise from left: empty shell casing, factory-made plastic wad, gunpowder, primer, shot. The size shot pictured is #6 — the best all-round size for general hunting.

Figure 7-4

Removing a primer from a fired shell. The shallow hole is 7/8" in diameter. The smaller hole is 3/8" in diameter.

Figure 7-5

Seating a new primer. Start the new primer by hand with finger pressure (upper left). Finish seating the primer by inserting a wooden dowel into the shell and applying simultaneous pressure (with a vise) to the front end of the dowel and the rear end of the shell.

Figure 7-6

Add gunpowder to the shell with a funnel and a measuring spoon. Aluminum and plastic utensils are preferred because they are non-sparking.

Figure 7-7
Putting the wad in the shell is a problem because of the crimp. The crimp can be spread, however, by inserting and twisting a tapered wooden dowel (far left). In sequence to the right is shown a commercial plastic wad; a wad inserted into the mouth, previously spread, of a shell casing; and a pencil being used to push down and seat a wad within the shell casing.

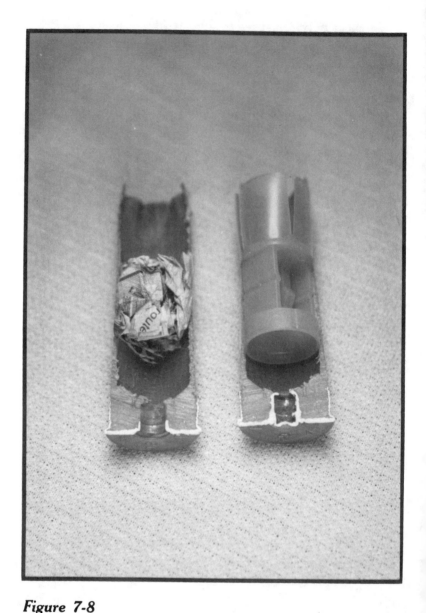

Figure 7-8

In this sectional view of a shotgun shell, on the left is a homemade wad of paper, on the right is a factory-made one-piece plastic wad and shot cup.

Recrimping the top of the shell with finger pressure only is difficult but can be done. Figure 7-9 shows actual shells which have been crimped with finger pressure alone. If the crimp simply won't stay in place but keeps unfolding, the mouth of the shell can be plugged up with a small ball of paper. The paper will prevent the shot from leaking out.

Figure 7-9

On the right is a plastic shotshell that has been recrimped totally by hand with finger pressure only. On the left is a shell which refused to stay crimped. A small ball of wadded up paper has been inserted as an overshot wad to stop the shot from leaking out. The crimp will hold the wad in place and the wad will hold the shot in place.

Patterns

To check your shotshell load, fire at a large sheet of paper 20 yards away. A man's stride is about equal to a yard. Draw a circle 30" in diameter around the most dense concentration of shot, count the shot particles inside the circle, and compare it to the number of shot particles in the original load.

A factory-made "cylinder bore" shotgun will place 35% of its pellets in a 30" diameter circle *at 40 yards*. Your homemade gun and homemade ammo won't do that well. That's why I suggest patterning your gun at only 20 yards. (In general hunting, that's the distance at which most of your shots will occur anyway.) By patterning, I mean experimenting with different loads — kinds and quantities of powder, kinds and quantities of shot, kinds of wads, etc. —to achieve a load that will place the highest percentage of shot inside the circle.

A common error is to load too much gunpowder in an effort to get a harder hitting load with longer range. The result often is that it "blows a hole" in the pattern. Experiment, of course, but always with a "safety first" attitude! Good luck and good shooting!

BIBLIOGRAPHY

Abrusci, Joseph, *Professional Homemade Salutes,* Cornville, AZ: Desert Publications, 1979.

Ahern, Jerry, "The Gun — Ultimate Symbol of Self-Reliance" *Guns & Ammo*, June 1980.

Avery, Ralph, *Combat Loads For The Sniper Rifle*, Cornville, AZ: Desert Publications, 1981.

Bebie, Jules, *Manual Of Explosives, Military Pyrotechnics, And Chemical Warfare Agents*, Boulder, CO: Paladin Press, 1942.

Benson, Ragnar, *Live Off The Land In The City And Country*, Boulder, CO: Paladin Press, 1982.

Benson, Ragnar, *Survival Poaching*, Boulder, CO: Paladin Press, 1980.

Boddington, Craig, "Big Game at Long Range" *Magnum Rifles* compiled by *Guns & Ammo* Specialty Books Staff, 1980.

Brown, Ronald, "Strike Anywhere Ammo" *Survive* (now *Guns & Action*) January/February 1983.

CIA Field Expedient Methods For Explosives Preparations, Cornville, AZ: Desert Publications, 1977.

179

CIA Field Expedient Preparation of Black Powders, Cornville, AZ: Desert Publications, 1977.

Danisevich, Philip J., *High-Low Boom!,* McDonald, OH: Ken Hale Publications, 1966.

Davis, Tenney L., *The Chemistry Of Powder And Explosives,* Hollywood, CA: Angriff Press, 1943.

Elias, Stephen, *Legal Research,* Berkeley, CA: Nolo Press, 1983.

Gottlieb, Alan M., *The Rights Of Gun Owners,* Aurora, IL: Caroline House Publishers, Inc., 1981.

Encyclopedia Americana, 1919 "Electrolysis," "Gunpowder," "Potassium Chlorate," "Potassium Nitrate".

"Explosives, Pyrotechnics, and Matches," *The Chemical Formulary* circa 1900.

Grennell, Dean, *The ABC's of Reloading,* Northfield, IL: DBI Books, 1974.

Gruener, Hippolyte, *The Story of Chemistry,* New York: P.F. Collier & Son, 1939.

Hacker, Rick, "Loading Your Muzzleloaders" Hunting Guns, *Outdoor Life,* 1984.

Hamilton, Claud, "A Hard Look at Handgun Brass" *Handloader Number 84* March-April 1980.

Hiscox, Gardner D., *Henley's Formulas For Home And Workshop,* New York: Avenel Books, 1979.

Hogg, Ivan V., *Guns And How They Work,* New York: Everest House, 1979.

Holmes, Bill, *Home Workshop Guns For Defense And Resistance — Volume One: The Submachine Gun,* Boulder, CO: Paladin Press, 1977.

Improvised Explosives For Use In Detonators, Cornville, AZ: Desert Pubications, 1980.

Improvised Munitions Black Book, Volume 3, Cornville, AZ: Desert Publications, 1982.

Introduction To Pyrotechnics, Nashua, NH: Pioneer Industries.

Kirkland, Turner, *Dixie Gun Works, 1981 Catalog,* Union City, TN: Dixie Gun Works, Inc.

Lee Load-All Junior Tables, Hartford, WI: Lee Precision, Inc., 1978.

Lewis, Tim, **Kitchen Improvised Plastic Explosives,** Odessa, TX: Information Publishing Co., 1983.

Luzadder, Warren, **Basic Graphics,** Englewood Cliffs, NJ: Prentice-Hall, 1964.

Markham & Smith, **General Chemistry,** Boston, MA: Houghton Mifflin Co., 1955.

Matunas, Ed, "All-Purpose 12-Gauge" *Handloader Number 93* September-October 1981.

Minnery, John, **How To Kill, Volume V,** Boulder, CO: Paladin Press, 1980.

Most, Johann, **Military Science For Revolutionaries,** Cornville, AZ: Desert Publications, 1978.

Oberg, Eric, et al. **Machinery's Handbook 22nd Edition,** New York: Industrial Press, 1984.

O'Connor, et al. **Chemistry Experiments And Principles,** Lexington, MA: D.C. Heath and Co., 1977.

Petzal, David E., **The Expert's Book Of The Shooting Arts,** New York: Simon and Schuster, 1972.

Powell, William, **The Anarchist Cookbook,** Secaucus, NJ: Lyle Stuart, Inc., 1979.

Reloaders' Guide for Hercules Smokeless Powders, Wilmington, DE: Hercules, Inc., 1978.

Remington Sporting Firearms and Ammunition for 1981 (a complimentary catalog) Bridgeport, CT: Remington Dupont.

Rice, F. Philip, **Gun Data Book,** New York: Harper & Row, 1975.

Saxon, Kurt, *The Poor Man's James Bond,* Harrison, AR: Atlan Formularies, 1972.

Serven, James E., *The Collecting Of Guns,* New York: Bonanza Books, 1964.

Shooting and Hunting Accessories 1980-81 (catalog) No. 32-B.

The Shooters Bible #41, New York: Stoeger Arms Corporation, 1950.

Speer Reloading Manual Number Ten, Lewiston, ID: Omark Industries, 1979.

Tappan, Mel, *Survival Guns,* Rogue River, OR: Janus Press, 1980.

US Department of Defense, *Emergency War Surgery,* Washington: US Government Printing Office, 1975.

US Army, *Improvised Munitions Handbook TM 31-210,* Philadelphia: Frankford Arsenal, 1969.

Waters, Ken, "American Bulleted Cartridges" *Gun Digest 20th Anniversary Edition,* Chicago: The Gun Digest Company, 1966.

"Weapons and Firearms" *American Jurisprudence 2nd Edition Volume 79,* Rochester, NY: The Lawyers Co-operative Publishing Co.

Weingart, George W., *Pyrotechnics,* 1947.

Winchester Western 1981 Sporting Arms and Ammunition (complimentary catalog), Olin Corp., 1980.

Zutz, Don, *Handloading For Hunters,* Tulsa, OK: Winchester Press, 1977.

Zwirz, Bob, "Recoil: Its Causes, Effects, and Remedies" *Ammo* August 1980.

OTHER BOOKS OF INTEREST:

ARMED DEFENSE, Gunfight Survival For The Householder and Businessman, *by Burt Rapp.* This book will show you how to teach yourself to shoot well enough to save your life in variety of ugly situations. You will learn techniques and tactics that work, not just reflections of somebody's theories. This book also covers what you need to know about the legal and emotional aspects of surviving a gunfight. *1989, 5½ x 8½, 214 pp, illustrated, soft cover.* **$16.95.**

PERSONAL DEFENSE WEAPONS, *by J. Randall.* The author, a private detective and weapons buff, evaluates all kinds of weapons — guns, knives, sticks, gas canisters, martial arts weapons, and many others — by asking some very interesting questions: Is it too deadly to use? Is it illegal to carry? Can it be comfortably concealed? How much skill does it take? Will it gross you out to use it? Is it reliable? Whatever you situation, this practical book will help you find protection you can live with. *1992, 5½ x 8½, 102 pp, illustrated, soft cover.* **$12.00.**

CLOSE SHAVES, The Complete Book of Razor Fighting, *by Bradley J. Steiner.* This is the *only* book ever written on the bloody art of razor fighting. More than 60 photographs demonstrate the easy-to-understand techniques. Carrying and Concealing razors; Fighting Grips; Stance and Body Movement; Vital Target Areas; Basic Attacks; How to Win a Razor Fight; And Much, Much More. WARNING: This book is pretty gruesome. *1980, 5½ x 8½, 86 pp, illustrated, soft cover.* **$10.00.**

You can get these books at your favorite bookstore or contact any of our distributors:

Bookpeople
7900 Edgewater Drive
Oakland, CA 94261
1-800-999-4650

Homestead Books
6101 22nd Avenue NW
Seattle, WA 98107
1-800-426-6777

Ingram Book Company
One Ingram Blvd.
La Vergne, TN 37086-1986
1-800-937-8000

Last Gasp of San Francisco
2948 20th Street
San Francisco, CA 94110
1-415-824-6636

Left Bank Distribution
1404 18th Avenue
Seattle, WA 98122
1-206-322-2868

Loompanics Unlimited
PO Box 1197
Port Townsend, WA 98368
1-800-380-2230

Marginal Distribution
277 George Street N Unit 102
Peterborough, Ontario
Canada K9J 3G9
1-705-745-2326

Van Patten Publishing
19741 41st Avenue NE
Seattle, WA 98155
1-206-306-7187